November 2013

INFORMATION TECHNOLOGY

Additional OMB and Agency Actions Are Needed to Achieve Portfolio Savings

INFORMATION TECHNOLOGY

Additional OMB and Agency Actions Are Needed to Achieve Portfolio Savings

Highlights of GAO-14-65, a report to congressional requesters

Why GAO Did This Study

Federal agencies plan to spend at least $82 billion on IT in fiscal year 2014, and GAO has previously reported on challenges in identifying and reducing duplicative IT investments. In 2012, OMB launched its PortfolioStat initiative—a process where agencies gather information on their IT investments and develop plans for consolidation and increased use of shared-service delivery models. GAO was asked to review the implementation of PortfolioStat. GAO's objectives were to (1) determine whether agencies completed key required PortfolioStat actions, (2) evaluate selected agencies' plans for making portfolio improvements and achieving associated cost savings, and (3) evaluate OMB's plans to improve the PortfolioStat process. To do this, GAO analyzed plans, status reports, and other documentation from agencies and interviewed agency and OMB officials. GAO also interviewed officials and reviewed documentation from five agencies selected based on their IT expenditures and management structures.

What GAO Recommends

GAO is recommending, among other things, that OMB require agencies to fully disclose limitations in CIOs' ability to exercise their authority and that 24 agencies take steps to improve their PortfolioStat implementation. OMB agreed with some of the recommendations and disagreed with others; and responses from the 21 agencies commenting on the report varied. GAO continues to believe that the majority of the recommendations are valid, but has removed two, and modified one, as discussed in the report.

View GAO-14-65. For more information, contact David A. Powner at (202) 512-9286 or pownerd@gao.gov.

What GAO Found

The 26 major federal agencies that were required to participate in the PortfolioStat initiative fully addressed four of seven key requirements established by the Office of Management and Budget (OMB). However, only 1 of the 26 agencies addressed all the requirements. For example, agencies did not develop action plans that addressed all elements, such as criteria for identifying wasteful, low-value or duplicative information technology (IT) investments, or migrate two commodity IT areas—such as enterprise IT systems and IT infrastructure—to a shared service by the end of 2012. (See table.)

Number of 26 Agencies Meeting PortfolioStat Requirements

Designate a lead official	Complete portfolio survey	Develop commodity IT baseline	Hold PortfolioStat session	Complete action plan	Migrate two services	Submit lessons learned
26	26	14	26	4	13	26

Source: GAO analysis of agency data.

In addition, some of these agencies had weaknesses in selected areas:

- 6 agencies reported limitations in their chief information officer's (CIO) authority to review and approve the entire portfolio;
- 5 agencies did not include all their investments in their enterprise architecture (i.e., an organizational blueprint), limiting their ability to identify investments to be consolidated or eliminated; and
- 12 agencies reported challenges in ensuring the completeness of their commodity IT baseline data or did not identify a process to ensure its completeness.

Further, OMB's estimate of about 100 consolidation opportunities and a potential $2.5 billion in savings from the PortfolioStat initiative is understated because, among other things, it did not include estimates from the Departments of Defense and Justice. GAO's analysis, which includes these estimates, shows that, collectively, the 26 agencies are reporting about 200 opportunities and at least $5.8 billion in potential savings through fiscal year 2015.

Five selected agencies—the Departments of Agriculture, Defense, the Interior, the Treasury, and Veterans Affairs—identified 52 consolidation initiatives, along with other IT management improvements, and estimated at least $3.7 billion in potential cost savings through fiscal year 2015. However, four agencies did not always provide sufficient support for all of their estimates, and they varied in their use of processes—such as an enterprise architecture and a method for assessing the value of investments—recommended by OMB to identify consolidation opportunities. More consistently using these tools may reveal further opportunities for consolidation, and better support for estimated savings may increase the chances that they will be achieved.

OMB's fiscal year 2013 PortfolioStat guidance identifies a number of planned improvements but does not fully address certain weaknesses in the implementation of the initiative, such as limitations in CIOs' authority, weaknesses in agencies' commodity IT baselines, accountability for migrating selected commodity IT areas, or the information on agencies' progress that OMB intends to make public.

_____ United States Government Accountability Office

Contents

Tables

Figure

Abbreviations

Agriculture	U.S. Department of Agriculture
CIO	Chief Information Officer
COO	Chief Operating Officer
Defense	Department of Defense
DHS	Department of Homeland Security
EA	enterprise architecture
EPA	Environmental Protection Agency
GSA	General Services Administration
HHS	Department of Health and Human Services
HUD	Department of Housing and Urban Development
Interior	Department of the Interior
IT	information technology
NARA	National Archives and Records Administration
NASA	National Aeronautics and Space Administration
NRC	U.S. Nuclear Regulatory Commission
NSF	National Science Foundation
OMB	Office of Management and Budget
OPM	Office of Personnel Management
SBA	Small Business Administration
SSA	Social Security Administration
Treasury	Department of the Treasury
USACE	U.S. Army Corps of Engineers
USAID	U.S. Agency for International Development
VA	Department of Veterans Affairs

GAO U.S. GOVERNMENT ACCOUNTABILITY OFFICE

441 G St. N.W.
Washington, DC 20548

November 6, 2013

The Honorable Thomas R. Carper
Chairman
The Honorable Tom Coburn, M.D.
Ranking Member
Committee on Homeland Security and Governmental Affairs
United States Senate

The Honorable Claire McCaskill
Chairman
Subcommittee on Financial and Contracting Oversight
Committee on Homeland Security and Governmental Affairs
United States Senate

The Honorable Susan M. Collins
United States Senate

Federal agencies expect to spend about $82 billion in fiscal year 2014 to meet their increasing demand for information technology (IT). Given the proliferation of duplicative, wasteful, low-value investments that both we and the Office of Management and Budget (OMB) have highlighted over the years, it is important that federal agencies avoid investing in these types of investments whenever possible. Avoiding such investments will be particularly important to ensure the most efficient use of resources as agencies continue to be faced with budget restrictions.

Over the past few years, we have issued a series of reports that have identified federal programs or functional areas where potentially unnecessary duplication, overlap, or fragmentation exists, the actions needed to address such conditions, and the potential financial and other benefits of doing so.[1] In March 2012, OMB launched an initiative, referred to as PortfolioStat, which requires agencies to conduct annual reviews of

---

[1]GAO, *2013 Annual Report: Actions Needed to Reduce Fragmentation, Overlap and Duplication, and Achieve Other Financial Benefits*, GAO-13-279SP (Washington, D.C.: Apr. 9, 2013); *2012 Annual Report: Opportunities to Reduce Duplication, Overlap and Fragmentation, Achieve Savings, and Enhance Revenue*, GAO-12-342SP (Washington, D.C.: Feb. 28, 2012); and *Opportunities to Reduce Potential Duplication in Government Programs, Save Tax Dollars, and Enhance Revenue*, GAO-11-318SP (Washington, D.C.: Mar. 1, 2011).

their IT investments and make decisions on eliminating duplication, among other things. According to OMB, PortfolioStat has the potential to save the government $2.5 billion over the next 3 years. Given the potential for achieving such savings and improving IT investment management, you asked us to review the implementation of PortfolioStat. Our specific objectives were to (1) determine the status of efforts to implement key required PortfolioStat actions, (2) evaluate selected agencies' plans for making portfolio improvements and achieving associated cost savings, and (3) evaluate OMB's plans to improve the PortfolioStat process.

To address our first objective, we obtained documentation from the 26 agencies[2] that were required to comply with OMB's memo for implementing the PortfolioStat initiative and compared it to the memo's requirements and supporting guidance. To address our second objective, we selected 5 agencies based on criteria including total IT expenditures for fiscal year 2012, investment management maturity level, and IT and Chief Information Officer organizational structures. These agencies are the Departments of Agriculture, Defense, the Interior, the Treasury, and Veterans Affairs. We analyzed these agencies' documentation and interviewed relevant officials to identify their plans for reducing potentially duplicative, low-value, and wasteful commodity IT investments and determine the extent to which they were based on the use of an enterprise architecture (EA)[3] and an IT valuation model,[4] as recommended by OMB, and the extent to which support for estimated cost savings was documented. To address our third objective, we reviewed OMB's guidance for the 2013 PortfolioStat and interviewed

---

[2]The 26 agencies are the Departments of Agriculture, Commerce, Defense, Education, Energy, Health and Human Services, Homeland Security, Housing and Urban Development, the Interior, Justice, Labor, State, Transportation, the Treasury, and Veterans Affairs; the Environmental Protection Agency, General Services Administration, National Aeronautics and Space Administration, National Archives and Records Administration, National Science Foundation, Office of Personnel Management, Small Business Administration, Social Security Administration, U.S. Agency for International Development, U.S. Army Corps of Engineers and the U.S. Nuclear Regulatory Commission.

[3]An enterprise architecture is a modernization blueprint that describes the organization's current and desired state for business operations and supporting IT systems in both logical and technical terms, and contains a plan for transitioning between the two states.

[4]An IT valuation model or methodology is used to comparatively evaluate investments based on their value to the agency and the cost to the taxpayer using a set of agency-defined criteria.

OMB's PortfolioStat Lead regarding plans for improving the PortfolioStat process. In addition, we analyzed the information obtained from our sources and the results of our analyses for our first two objectives to determine whether OMB's plans for improving PortfolioStat addressed the issues we identified.

We conducted this performance audit from October 2012 to November 2013 in accordance with generally accepted government auditing standards. Those standards require that we plan and perform the audit to obtain sufficient, appropriate evidence to provide a reasonable basis for our findings and conclusions based on our audit objectives. We believe that the evidence obtained provides a reasonable basis for our findings and conclusions based on our audit objectives. Appendix I contains additional details on our objectives, scope, and methodology.

Background

Information technology should enable government to better serve the American people. However, OMB stated in 2010 that the federal government had achieved little of the productivity improvements that private industry had realized from IT, despite spending more than $600 billion on IT over the past decade.[5] Too often, federal IT projects run over budget, behind schedule, or fail to deliver promised functionality.[6]

Both OMB and federal agencies have key roles and responsibilities for overseeing IT investment management. OMB is responsible for working with agencies to ensure investments are appropriately planned and justified.[7] Federal agencies are responsible for managing their IT investment portfolio, including the risks from their major information system initiatives, in order to maximize the value of these investments to the agency. Additionally, each year, OMB and federal agencies work together to determine how much the government plans to spend on IT

---

[5]OMB, *25 Point Implementation Plan to Reform Federal Information Technology Management* (Washington, D.C.: December 2010).

[6]See GAO, *Information Technology: OMB and Agencies Need to More Effectively Implement Major Initiatives to Save Billions of Dollars*, GAO-13-796T (Washington, D.C.: July 25, 2013).

[7]The Clinger-Cohen Act sets out requirements for capital planning and investment control and for performance-based and results-based management. 40 U.S.C. §§ 11302-11303.

projects and how these funds are to be allocated. For fiscal year 2014, federal agencies plan to spend about $82 billion.

GAO Has Previously Reported on Opportunities to Reduce Duplication and Achieve Cost Savings in Critical IT-Related Areas

We have previously reported on the challenges associated with agencies' efforts to identify duplicative IT investments. For example, in September 2011 we reported that there were hundreds of investments providing similar functions across the federal government, including 1,536 information and technology management investments, 781 supply chain management investments, and 661 human resource management investments.[8] Further, we found that OMB guidance to agencies on how to report their IT investments did not ensure complete reporting or facilitate the identification of duplicative investments. Specifically, agencies differed on what investments they included as an IT investment, and OMB's guidance requires each investment to be mapped to a single functional category. As a result, agencies' annual IT investments were likely greater that the $79 billion reported in fiscal year 2011 and OMB's ability to identify duplicative investments was limited. Further, we found that several agencies did not routinely assess operational systems to determine if they were duplicative. We recommended, among other things, that OMB clarify its guidance to help agencies better identify and categorize their IT investments and require agencies to report the steps they take to ensure that their IT investments are not duplicative. OMB agreed with these recommendations.

More recently, we reported on efforts at the Departments of Defense, Energy, and Homeland Security to identify duplicative IT investments.[9] More specifically, we noted that although Defense, Energy, and Homeland Security use various investment review processes to identify duplicative investments, 37 of our sample of 810 investments were potentially duplicative at Defense and Energy. These investments accounted for about $1.2 billion in spending for fiscal years 2007 through 2012. We also noted that investments were, in certain cases, misclassified by function, further complicating agencies' ability to identify and eliminate duplicative investments. We recommended that Defense

---

[8]GAO, *Information Technology: OMB Needs to Improve Its Guidance on IT Investments*, GAO-11-826 (Washington, D.C.: Sept. 29, 2011).

[9]GAO, *Information Technology: Departments of Defense and Energy Need to Address Potentially Duplicative Investments*, GAO-12-241 (Washington, D.C.: Feb. 17, 2012).

GAO-14-65 OMB PortfolioStat Initiative

and Energy utilize transparency mechanisms, such as the IT Dashboard[10] to report on the results of their efforts to identify and eliminate potentially duplicative investments. The agencies generally agreed with this recommendation.

We have also reported on the value of portfolio management in helping to identify duplication and overlap and opportunities to leverage shared services. For example, we have reported extensively on various agencies' IT investment management capabilities by using GAO's IT Investment Management Framework,[11] in which stage 3 identifies best practices for portfolio management, including (1) creating a portfolio which involves, among other things, grouping investments and proposals into predefined logical categories so they can be compared to one another within and across the portfolio categories, and the best overall portfolio can then be selected for funding, and (2) evaluating the portfolio by monitoring and controlling it to ensure it provides the maximum benefits at a desired cost and an acceptable level of risk.

OMB Established PortfolioStat to Help Agencies Reduce Duplication and Achieve Cost Savings

Recognizing the proliferation of duplicative and low-priority IT investments within the federal government and the need to drive efficiency, OMB launched the PortfolioStat initiative in March 2012, which requires 26 agencies to conduct an annual agency-wide IT portfolio review to, among other things, reduce commodity IT spending and demonstrate how their IT investments align with the agency's mission and business functions.[12] Toward this end, OMB defined 13 types of commodity IT investments in three broad categories:

(1) Enterprise IT systems, which include e-mail; identity and access management; IT security; web hosting, infrastructure, and content; and collaboration tools.

---

[10]In June 2009, OMB deployed a public website known as the IT Dashboard, which provides detailed information on federal agencies' major IT investments, including assessments of actual performance against cost and schedule targets (referred to as ratings).

[11]GAO, *Information Technology Investment Management: A Framework for Assessing and Improving Process Maturity (Version 1.1)*, GAO-04-394G (Washington, D.C.: March 2004).

[12]OMB, *Implementing PortfolioStat*, Memorandum M-12-10 (Washington, D.C.: Mar. 30, 2012).

(2) IT infrastructure, which includes desktop systems, mainframes and servers, mobile devices, and telecommunications.

(3) Business systems, which include financial management, grants-related federal financial assistance, grants-related transfer to state and local governments, and human resources management systems.[13]

PortfolioStat is designed to assist agencies in assessing the current maturity of their IT investment management process, making decisions on eliminating duplicative investments, and moving to shared solutions (such as cloud computing[14]) in order to maximize the return on IT investments across the portfolio. It is also intended to assist agencies in meeting the targets and requirements under other OMB initiatives aimed at eliminating waste and duplication and promoting shared services, such as the Federal Data Center Consolidation Initiative, the Cloud Computing Initiative, and the IT Shared Services Strategy.

PortfolioStat is structured around five phases: (1) *baseline data gathering* in which agencies are required to complete a high-level survey of their IT portfolio status and establish a commodity IT baseline; (2) *analysis and proposed action plan* in which agencies are to use the data gathered in phase 1 and other available agency data to develop a proposed action plan for consolidating commodity IT; (3) *PortfolioStat session* in which agencies are required to hold a face-to-face, evidence-based review of their IT portfolio with the Federal Chief Information Officer (CIO) and key stakeholders from the agency to discuss the agency's portfolio data and proposed action plan, and agree on concrete next steps to rationalize the agency's IT portfolio that would result in a final plan; (4) *final action plan implementation*, in which agencies are to, among other things, migrate at least two commodity IT investments; and (5) *lessons learned*, in which agencies are required to document lessons learned, successes, and

---

[13]OMB made available "Frequently Asked Questions" which included definitions of these types of commodity IT investment based on OMB Circular A-11 and the Federal Enterprise Architecture Consolidated Reference Model (OMB, *Circular No. A-11, Preparation, Submission, and Execution of the Budget* (Washington, D.C.: August 2012); OMB, *FEA Consolidated Reference Model Document Version 2.3* (Washington, D.C.: October 2007)).

[14]Cloud computing is an emerging form of delivering computing services via networks with the potential to provide IT services more quickly and at a lower cost. Cloud computing provides users with on-demand access to a shared and scalable pool of computing resources with minimal management effort or service provider interaction.

challenges. Each of these phases is associated with more specific requirements and deadlines.

OMB has reported that the PortfolioStat effort has the potential to save the government $2.5 billion through fiscal year 2015 by consolidating and eliminating duplicative systems.

Agencies Addressed PortfolioStat Requirements, but Baselines and Consolidation Plans Were Not All Complete

In its memo on implementing PortfolioStat, OMB established several key requirements for agencies: (1) designating a lead official with responsibility for implementing the process; (2) completing a high-level survey of their IT portfolio; (3) developing a baseline of the number, types, and costs of their commodity IT investments; (4) holding a face-to-face PortfolioStat session with key stakeholders to agree on actions to address duplication and inefficiencies in their commodity IT investments; (5) developing final action plans to document these actions; (6) migrating two commodity IT areas to shared services; and (7) documenting lessons learned.[15] In addition, in guidance supporting the memo, agencies were asked to report estimated savings and cost avoidance associated with their consolidation and shared service initiatives through fiscal year 2015.

All 26 agencies that were required to implement the PortfolioStat process took actions to address OMB's requirements. However, there were shortcomings in their implementation of selected requirements, such as addressing all required elements of the final action plan and migrating two commodity areas to a shared service by the end of 2012. Table 1 summarizes the agencies' implementation of the requirements in the memo, which are discussed in more detail below.

---

[15]OMB M-12-10.

Table 1: Agencies' Implementation of PortfolioStat Requirements

Agency	Designate PortfolioStat lead	Complete IT portfolio survey	Develop commodity IT baseline[a]	Hold PortfolioStat session	Complete action plan with required elements	Complete two migration efforts	Submit lessons learned	Requirements met
Agriculture	Yes	Yes	No	Yes	No	Yes	Yes	5 of 7
Commerce	Yes	Yes	No	Yes	Yes	Yes	Yes	6 of 7
Defense	Yes	Yes	No	Yes	No	Yes	Yes[b]	5 of 7
DHS	Yes	Yes	Yes	Yes	No	Yes	Yes	6 of 7
Education	Yes	Yes	Yes	Yes	Yes	Yes	Yes	7 of 7
Energy	Yes	Yes	Yes	Yes	No	Yes	Yes	6 of 7
EPA	Yes	Yes	No	Yes	No	No	Yes	4 of 7
GSA	Yes	Yes	Yes	Yes	Yes	No	Yes	6 of 7
HHS	Yes	Yes	Yes	Yes	No	Yes	Yes	6 of 7
HUD	Yes	Yes	No	Yes	No	No	Yes[b]	4 of 7
Interior	Yes	Yes	No	Yes	No	No	Yes	4 of 7
Justice	Yes	Yes	Yes	Yes	No	Yes	Yes	6 of 7
Labor	Yes	Yes	No	Yes	No	No	Yes	4 of 7
NARA	Yes	Yes	Yes	Yes	No	Yes	Yes	6 of 7
NASA	Yes	Yes	Yes	Yes	No	No	Yes	5 of 7
NRC	Yes	Yes	No	Yes	No	Yes	Yes	5 of 7
NSF	Yes	Yes	Yes	Yes	No	Yes	Yes	6 of 7
OPM	Yes	Yes	No	Yes	No	No	Yes	4 of 7
SBA	Yes	Yes	No	Yes	No	Yes	Yes	5 of 7
SSA	Yes	Yes	No	Yes	Yes	No	Yes	5 of 7
State	Yes	Yes	Yes	Yes	No	No	Yes[b]	5 of 7
Transportation	Yes	Yes	Yes	Yes	No	No	Yes	5 of 7
Treasury	Yes	Yes	Yes	Yes	No	Yes	Yes	6 of 7
USACE	Yes	Yes	Yes	Yes	No	No	Yes	5 of 7
USAID	Yes	Yes	No	Yes	No	No	Yes	4 of 7
VA	Yes	Yes	Yes	Yes	No	No	Yes	5 of 7
No. of agencies that met requirement	**26 of 26**	**26 of 26**	**14 of 26**	**26 of 26**	**4 of 26**	**13 of 26**	**26 of 26**	

Source: GAO analysis of agency documentation.

Note: DHS—Department of Homeland Security; EPA—Environmental Protection Agency; GSA—General Services Administration; HHS—Department of Health and Human Services; HUD—Department of Housing and Urban Development; NARA—National Archives and Records Administration; NASA—National Aeronautics and Space Administration; NRC—U.S. Nuclear Regulatory Commission; NSF—National Science Foundation; OPM—Office of Personnel Management; SBA—Small Business Administration; SSA—Social Security Administration; USACE—

U.S. Army Corps of Engineers; USAID—U.S. Agency for International Development; VA—Department of Veterans Affairs.

[a]Agencies were given a rating of "no" if they could not ensure the completeness of their baseline information.

[b]Agencies reported that they had submitted lessons learned information in their final action plan.

Agencies Generally Designated the CIO as the Lead for PortfolioStat Efforts

In the memo for implementing the PortfolioStat initiative, OMB required each agency's chief operating officer (COO) to designate and communicate within 10 days of the issuance of the memo an individual with direct reporting authority to the COO to lead the agency's PortfolioStat implementation efforts. Consistent with a recent OMB memo requiring chief information officers (CIO) to take responsibility for commodity IT,[16] 19 of the 26 agencies designated the CIO or chief technology officer to lead their PortfolioStat efforts. The remaining 7 agencies designated the Assistant Attorney General for Administration (Department of Justice), the deputy CIO (Department of Transportation), the Assistant Secretary for Management (Department of the Treasury), the Office of Information and Technology Chief Financial Officer (Department of Veterans Affairs), the Director, Office of Information Resources Management, Chief Human Capital Officer (National Science Foundation), the Senior Advisor to the Deputy Commissioner/Chief Operating Officer (Social Security Administration), and the Senior Deputy Assistant Administrator (U.S. Agency for International Development).

Portfolio Survey Provided Status of CIO Authority and Other Issues

As part of the baseline data-gathering phase, OMB required agencies to complete a high-level survey of the status of their IT portfolio. The survey asked agencies to provide information related to implementing OMB guidance, including information on the CIO's explicit authority to review and approve the entire IT portfolio,[17] the percentage of IT investments that are reflected in the agency's EA (required in OMB Circular A-130),[18] and the percentage of agency IT investments (major and non-major) that

---

[16]OMB, *Memorandum for Heads of Executive Departments and Agencies: Chief Information Officer Authorities*, M-11-29 (Washington, D.C.: Aug. 8, 2011).

[17]OMB M-11-29 states that CIOs must drive the investment review process for IT investments and have responsibility over the entire IT portfolio for an agency.

[18]OMB, *Management of Federal Information Resources,* Circular A-130 (Washington, D.C.: Nov. 28, 2000).

have gone through the TechStat process, both agency-led and OMB-led (required in OMB M-11-29).[19]

While all 26 agencies completed the survey, the survey responses highlighted that agencies varied in the maturity of their IT portfolio management practices. In particular, 6 agencies reported varying levels of CIO authority, 5 agencies reported that less than 100 percent of investments were reflected in the agency's EA, and most agencies noted that less than 50 percent of their major and non-major investments had gone through the TechStat process.

Following are highlights of their responses:

CIO authority: Twenty of the 26 agencies stated that they either had a formal memorandum or policy in place explicitly noting the CIO's authority to review and approve the entire agency IT portfolio or that the CIO collaborated with others (such as members of an investment review board) to exercise this authority. However, the remaining 6 agencies either reported that the CIO did not have this authority or there were limitations to the CIO's authority:

- The Department of Energy reported that while its CIO worked with IT governance groups, by law, the department CIO has no direct authority over IT investments in two semi-autonomous agencies (the National Nuclear Security Administration and the Energy Information Administration[20]).

- Although the Department of Health and Human Services reported having a formal memo in place outlining the CIO's authority and ability to review the entire IT portfolio, it also noted that the CIO had limited influence and ability to recommend changes to it.

---

[19]In January 2010, OMB began conducting TechStats, which are face-to-face, evidence-based reviews of an at-risk IT investment. Subsequently, as part of the Federal CIO's 25-point IT Reform Plan, OMB empowered agency CIOs to hold their own TechStat sessions within their respective agencies and required agencies to hold at least one TechStat session by March 2011, and one bureau-led TechStat review by June 2012. In August 2011, OMB M-11-29 required agency CIOs to continue holding TechStat sessions.

[20]The Department of Energy referred to a statutory basis to support its statement about two semi-autonomous agencies. See generally National Defense Authorization Act for Fiscal Year 2000, Pub. L. No. 106-65, title XXXII, 113 Stat. 512, 953 (1999); Department of Energy Organization Act, Pub. L. No. 95-91, 91 Stat. 565 (1977).

GAO-14-65 OMB PortfolioStat Initiative

- The Department of State reported that its CIO currently has authority over just 40 percent of IT investments within the department.

- The National Aeronautics and Space Administration reported that its CIO does not have authority to review and approve the entire agency IT portfolio.

- The Office of Personnel Management reported that the CIO advises the Director, who approves the IT portfolio, but this role is not explicitly defined.

- The U.S. Agency for International Development reported that the CIO's authority is limited to the portfolio that is executed within the office of the CIO.

It is important to note that OMB's survey did not specifically require agencies to disclose limitations their CIOs might have in their ability to exercise the authorities and responsibilities provided by law and OMB guidance. Thus it is not clear whether all those who have such limitations reported them or whether those who reported limitations disclosed all of them.

We recently reported that while federal law provides CIOs with adequate authority to manage IT for their agencies, limitations exist that impede their ability to exercise this authority.[21] We noted that OMB's memo on CIO authorities[22] was a positive step in reaffirming the importance of the role of CIOs in improving agency IT management, but did not require them to measure and report the progress of CIOs in carrying out these responsibilities. Consequently, we recommended that the Director of OMB establish deadlines and metrics that require agencies to demonstrate the extent to which their CIOs are exercising the authorities and responsibilities provided by law and OMB's guidance. In response, OMB stated that it would ask agencies to report on the implementation of the memo.

---

[21]GAO, *Federal Chief Information Officers: Opportunities Exist to Improve Role in Information Technology Management*, GAO-11-634 (Washington, D.C.: Sept. 15, 2011).

[22]OMB, *Memorandum for Heads of Executive Departments and Agencies: Chief Information Officer Authorities*, M-11-29 (Washington, D.C.: Aug. 8, 2011).

The high-level survey responses regarding CIO authority at agencies indicate that several CIOs still do not exercise the authority needed to review and approve the entire IT portfolio, consistent with OMB guidance. Although OMB has issued guidance and required agencies to report on actions taken to implement it, this has not been sufficient to ensure that agency COOs address the issue of CIO authority at their respective agencies. As a result, agencies are hindered in addressing certain responsibilities set out in the Clinger-Cohen Act of 1996,[23] which established the position of CIO to advise and assist agency heads in managing IT investments. Until the Director of OMB and the Federal CIO require agencies to fully disclose limitations their CIOs may have in exercising the authorities and responsibilities provided by law and OMB's guidance, OMB may lack crucial information needed to understand and address the factors that could prevent agencies' from successfully implementing the PortfolioStat initiative.

Investments reflected in agencies' enterprise architecture: Twenty one of the 26 agencies reported that 100 percent of their IT investments are reflected in their agency's EA, while the remaining 5 agencies reported less than 100 percent: Commerce (90 percent), Justice (97 percent), State (40 percent), National Aeronautics and Space Administration (17 percent), and U.S. Agency for International Development (75 percent). According to OMB guidance, agencies must support an architecture with a complete inventory of agency information resources, including stakeholders and customers, equipment, systems, services, and funds devoted to information resources management and IT, at an appropriate level of detail.[24] Until these agencies' enterprise architectures reflect 100 percent of their IT investments, they will be limited in their ability to use this tool as a mechanism to identify low-value, duplicative, or wasteful investments.

TechStat process: Twenty-one of the 26 agencies reported that less than 50 percent of major and non-major investments had gone through the TechStat process and 1 reported that more than 50 percent of its

---

[23]See Pub. L. No. 104-106, Div. E, 110 Stat. 186, 679 (1996); P.L. No. 104-208, 110 Stat. 3009, 3009-393 (1996); 40 U.S.C 11101, et seq.

[24]OMB, *Guidance on Exhibits 53 and 300 – Information Technology and E-Government* (Washington, D.C.: July 1, 2013).

investments had gone through the process.[25] As we have previously reported, TechStat accountability sessions have the value of focusing management attention on troubled projects and establishing clear action items to turn the projects around or terminate them.[26] In addition, the TechStat model is consistent with government and industry best practices for overseeing IT investments, including our own guidance on IT investment management processes.[27] Consistent with these survey responses, in June 2013 we reported that the number of TechStat sessions held to date was relatively small compared to the current number of medium- and high-risk IT investments at federal agencies. Accordingly, we recommended that OMB require agencies to conduct TechStat sessions on certain IT investments, depending on their risk level. Holding TechStat sessions will help strengthen overall IT governance and oversight and will help agencies to better manage their IT portfolio and reduce waste. OMB generally concurred with our recommendation and stated that it was taking steps to address it.

Commodity IT Baselines Were Not All Complete

As part of the baseline data-gathering phase, each of the 26 agencies was also required to develop a comprehensive commodity IT baseline including information on each of the 13 types of commodity IT. Among other things, they were to include the fiscal year 2011 obligations incurred for commodity IT services and the number of systems providing these services.

The 26 agencies reported that they obligated approximately $13.5 billion in fiscal year 2011 for commodity IT, with the majority of these obligations (about $8.1 billion) for investments related to IT Infrastructure. Agencies also classified approximately 71.2 percent of the commodity IT systems identified (about 1,937 of the 2,721 reported) as enterprise IT systems.[28] Further, as illustrated in figure 1, of the total systems reported, most were

---

[25]The remaining four agencies only addressed major investments in their response.

[26]GAO, *Information Technology: Additional Executive Review Sessions Needed to Address Troubled Projects*, GAO-13-524 (Washington, D.C.: June 13, 2013).

[27] GAO-04-394G.

[28]OMB did not request information on the number of systems associated with IT infrastructure.

related to IT security, whereas the fewest systems were related to grants-related transfer to state and local governments.

Figure 1: Number of Business and Enterprise Commodity IT Systems Reported by Agencies

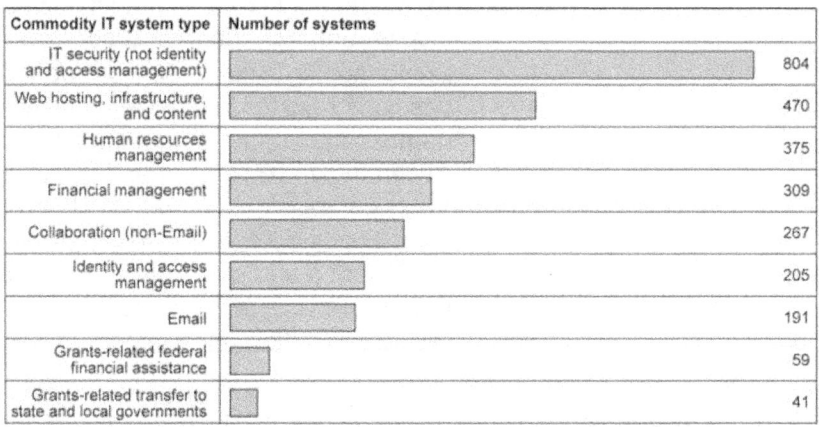

Commodity IT system type	Number of systems
IT security (not identity and access management)	804
Web hosting, infrastructure, and content	470
Human resources management	375
Financial management	309
Collaboration (non-Email)	267
Identity and access management	205
Email	191
Grants-related federal financial assistance	59
Grants-related transfer to state and local governments	41

Source: GAO analysis of agency and OMB reported data.

Note: The number of Defense business and enterprise commodity IT systems were not included because the department did not provide this information to OMB as part of the 2012 PortfolioStat.

When collecting data, it is important to have assurance that the data are accurate. We have previously reported on the need for agencies, when providing information to OMB, to explain the procedures used to verify their data.[29] Specifically, agencies should ensure that reported data are sufficiently complete, accurate, and consistent, and also identify any significant data limitations. Explaining the limitations of information can provide a context for understanding and assessing the challenges agencies face in gathering, processing, and analyzing needed data. We have also reiterated the importance of providing OMB with complete and

---

[29]GAO, *Data Center Consolidation: Agencies Need to Complete Inventories and Plans to Achieve Expected Savings*, GAO-11-565 (Washington, D.C.: July 19, 2011); and GAO, *Agencies' Annual Performance Plans under the Results Act: An Assessment Guide to Facilitate Congressional Decisionmaking*, GAO/GGD/AIMD-10.1.18 (Washington, D.C.: February 1998).

GAO-14-65 OMB PortfolioStat Initiative

accurate data and the possible negative impact of that data being missing or incomplete.[30]

While all 26 agencies developed commodity IT baselines, these baselines were not all complete. Specifically, 12 agencies (the Departments of Agriculture, Commerce, Defense, Housing and Urban Development, Labor, and the Interior; the Environmental Protection Agency, Nuclear Regulatory Commission, Office of Personnel Management, Small Business Administration, Social Security Administration, and U.S. Agency for International Development) could not ensure the completeness of their commodity IT baseline, either because they did not identify a process for this or faced challenges in collecting complete information. These agencies reported they were unable to ensure the completeness of their information for a range of reasons, including that they do not typically capture the required data at the level of detail required by OMB, that they used service contracts which do not allow visibility into specifics on the commodity IT inventory, that they lacked visibility into bureaus' commodity IT information, and that OMB's time frames did not allow for verification of information collected from lower-level units of the organization. Until agencies develop a complete commodity IT baseline, they may not have sufficient information to identify further consolidation opportunities.

While it is important that reported data are sufficiently complete, accurate, and consistent, OMB did not require agencies to verify their data or disclose any limitations on the data provided and does not plan to collect this information as agencies provide updated information in quarterly reporting. Until OMB requires agencies to verify their data and disclose any limitations in integrated data collection quarterly reporting, it may lack information it needs to more effectively oversee agencies' investment in commodity IT and identify Portfolio cost savings.

Key Stakeholders Generally Attended PortfolioStat Sessions

All 26 agencies held a PortfolioStat session in 2012, consistent with OMB's requirement. In addition, the agencies noted that the agency CIO, Chief Administrative Officer, Chief Financial Officer, and COO—the key stakeholders identified in OMB memorandum 12-10—in many instances

---

[30]GAO, *Information Technology: OMB Has Made Improvements to Its Dashboard, but Further Work Is Needed by Agencies and OMB to Ensure Data Accuracy*, GAO-11-262 (Washington, D.C.: Mar. 15, 2011).

attended this session.[31] In the instances where key stakeholders did not attend, authorized representatives of those stakeholders generally attended in their place, according to agency officials. Involvement from key stakeholders in agencies' PortfolioStat sessions is critical to ensuring agencies are maximizing their efforts to successfully implement PortfolioStat.

Agencies' Action Plans Did Not Always Address All Required Elements

Agencies were required by OMB to complete a final action plan that addressed eight specific elements: (1) describe plans to consolidate authority over commodity IT spending under the agency CIO;[32] (2) establish specific targets and deadlines for commodity IT spending reductions; (3) outline plans to migrate at least two commodity IT areas to shared services by December 31, 2012; (4) target duplicative systems or contracts that support common business functions for consolidation; (5) illustrate how investments within the IT portfolio align with the agency's mission and business functions; (6) establish criteria for identifying wasteful, "low-value," or duplicative investments; (7) establish a process to identify these potential investments and a schedule for eliminating them from the portfolio; and (8) improve governance and program management using best practices and, where possible, benchmarks.

All 26 agencies completed an action plan as required by OMB, but the extent to which they addressed the required items varied. Specifically, 18 agencies fully addressed at least six of the eight required elements—with Commerce, Education, General Services Administration, and Social Security Administration addressing all of them—and the remaining 8 agencies fully addressed five requirements or fewer and either partially addressed or did not address others. The consolidation of commodity IT spending under the agency CIO and establishment of criteria for identifying low-value, wasteful, and duplicative investments were the

---

[31]The Department of Education was unable to confirm the individuals who attended its PortfolioStat session, and the Department of Health and Human Services was only able to confirm that its CIO was in attendance. In addition, the National Science Foundation and the Office of Personnel Management stated that they did not have a chief administrative officer at the time of the 2012 PortfolioStat session.

[32]We recognize that there may be instances where legal or other constraints upon the agency may limit the CIO's consolidation of authority over commodity IT spending. In this report, however, we assess agencies' completion of a final action plan, required by OMB, an element of which is a description of plans to consolidate authority over commodity IT spending under the agency CIO.

elements that were addressed the least (12 and 9 agencies respectively); and the alignment of investments to the agency's mission and improvement of governance and program management were addressed by all agencies. Table 2 shows the extent to which the 26 agencies addressed the required elements in their action plan.

Until agencies address the items that were required in the PortfolioStat action plan in future OMB reporting, they will not be in a position to fully realize the intended benefits of the PortfolioStat initiative.

Table 2: Extent to Which Agency Action Plans Addressed Required Elements

Agency	Consolidate commodity IT spending under the CIO	Establish targets and deadlines for commodity IT spending reductions	Migrate at least two commodity IT areas to shared services	Target duplicative systems or contracts for consolidation	Mission and business function alignment	Criteria for wasteful, low-value, or duplicative investments	Process to identify and schedule to eliminate wasteful, low-value, or duplicative investments	Improve governance and program management
Agriculture	No	Yes	Yes	Yes	Yes	Yes	Yes	Yes
Commerce	Yes	Yes	Yes	Yes	Yes	Yes	Yes	Yes
Defense	No	Yes	Yes	Yes	Yes	Yes	Yes	Yes
DHS	No[a]	Yes	Yes	Yes	Yes	Yes	Yes	Yes
Education	Yes	Yes	Yes	Yes	Yes	Yes	Yes	Yes
Energy	No	Yes	Yes	Yes	Yes	No	Yes	Yes
EPA	No	Partially	Yes	Yes	Yes	No	Yes	Yes
GSA	Yes	Yes	Yes	Yes	Yes	Yes	Yes	Yes
HHS	Partially	Yes	Yes	Yes	Yes	Yes	Yes	Yes
HUD	Yes	Yes	Yes	Yes	Yes	No	Yes	Yes
Interior	Yes	Yes	Yes	Yes	Yes	No	Yes	Yes
Justice	Yes	Partially	Yes	Yes	Yes	Yes	Yes	Yes
Labor	Partially	Partially	Yes	Yes	Yes	Yes	Yes	Yes
NARA	No	Yes	Yes	Partially	Yes	No	No	Yes
NASA	No	Yes	Yes	Partially	Yes	No	No	Yes
NRC	Partially	No	Yes	Partially	Yes	Yes	Partially	Yes
NSF	No	Yes	Yes	Yes	Yes	No	Yes	Yes
OPM	Yes	Yes	Partially	Partially	Yes	Yes	Yes	Yes
SBA	Partially	No	Yes	Partially	Yes	Yes	Partially	Yes
SSA	Yes	Yes	Yes	Yes	Yes	Yes	Yes	Yes
State	No	Partially	No	Partially	Yes	Yes	Partially	Yes

GAO-14-65 OMB PortfolioStat Initiative

Agency	Consolidate commodity IT spending under the CIO	Establish targets and deadlines for commodity IT spending reductions	Migrate at least two commodity IT areas to shared services	Target duplicative systems or contracts for consolidation	Mission and business function alignment	Criteria for wasteful, low-value, or duplicative investments	Process to identify and schedule to eliminate wasteful, low-value, or duplicative investments	Improve governance and program management
Transportation	No	Partially	Yes	Partially	Yes	Yes	Partially	Yes
Treasury	No	Yes	Yes	Yes	Yes	No	Yes	Yes
USACE	No	Yes	Yes	Partially	Yes	No	Partially	Yes
USAID	Yes	Yes	Yes	Partially	Yes	Yes	Partially	Yes
VA	Yes	Yes	Yes	Partially	Yes	Yes	Yes	Yes

Source: GAO analysis of agency documentation.

Notes: A "partially" rating was given if the plan addressed a portion but not all of the information required in the element. DHS—Department of Homeland Security; EPA—Environmental Protection Agency; GSA—General Services Administration; HHS—Department of Health and Human Services; HUD—Department of Housing and Urban Development; NARA—National Archives and Records Administration; NASA—National Aeronautics and Space Administration; NRC— U.S. Nuclear Regulatory Commission; NSF—National Science Foundation; OPM—Office of Personnel Management; SBA—Small Business Administration; SSA—Social Security Administration; USACE—U.S. Army Corps of Engineers; USAID—U.S. Agency for International Development; VA—Department of Veterans Affairs.

[a]While Homeland Security's final PortfolioStat action plan did not include this required element, the department provided information to OMB in May 2013 that addressed their effort to consolidate commodity IT spending under the CIO. Therefore, we did not make a recommendation to Homeland Security to require them to report on this element in future OMB reporting.

Agencies Did Not All Complete Required Migration Efforts

Memorandum 12-10 required the 26 agencies to complete the migration of the two commodity IT areas mentioned in their action plan to shared services by December 31, 2012 (see app. II for the list of migration efforts by agency). However, 13 of the 26 agencies (the Departments of Housing and Urban Development, the Interior, Labor, State, Transportation and Veterans Affairs; the Environmental Protection Agency, General Services Administration, National Aeronautics and Space Administration, Office of Personnel Management, Social Security Administration, U.S. Agency for International Development, and the U.S. Army Corps of Engineers) reported that they still had not completed the migration of these areas as of August 2013. These agencies reported several reasons for this, including delays in establishing contracts with vendors due to the current budget situation, and delays due to technical challenges.

While OMB has stated that this initial requirement to migrate two systems was to initiate consolidation activities at the agencies, and not necessarily an action which it was intending to track for compliance, tracking the progress of such efforts would help to ensure accountability for agencies' results and the continued progress of PortfolioStat. OMB's 2013 PortfolioStat memo includes a requirement for agencies to report quarterly on the status of consolidation efforts and the actual and planned cost savings and/or avoidances achieved or expected, but the guidance does not specify that agencies should report on the status of the two migration efforts initiated in 2012. Until agencies report on the progress in consolidating the two commodity IT areas to shared services and OMB requires them to report on the status of these two efforts in the integrated data collection quarterly reporting, agencies will be held less accountable for the results of all their PortfolioStat efforts.

Lessons Learned Highlight Importance of CIO Authority and Value Engineering

Memorandum 12-10 required agencies to document and catalogue successes, challenges, and lessons learned from the PortfolioStat process into a document which was to be submitted to OMB by February 1, 2013.[33] Of the 26 agencies required to implement the PortfolioStat process, 23 agencies submitted lessons learned documentation. The 3 agencies that did not submit lessons learned in the format requested by OMB indicated that they did not submit this documentation because lessons learned had already been included in their final action plans.

Several agencies identified lessons learned related to the CIO's authority and the use of an IT valuation model (12 and 15, respectively).[34] More specifically, 8 agencies noted that OMB's requirements for a plan to consolidate commodity IT spending under the agency CIO and to identify the extent to which the CIO possesses explicit agency authority to review and approve the entire agency IT portfolio had enabled their agencies to improve the management of their commodity IT and portfolio. Further, 4 agencies stated that the requirements regarding CIO authority would help

---

[33]OMB provided agencies a template for documenting successes, challenges, and lessons learned in key areas, including CIO authority, valuation model, and program management, although several agencies used their own format.

[34]An IT valuation model or methodology is used to comparatively evaluate investments based on their value to the agency and the cost to the taxpayer using a set of agency-defined criteria. The federal CIO Council Best Practices committee issued a how-to-guide in October 2002 whose purpose is to define, capture, and measure value associated with electronic services unaccounted for in traditional return-on-investment calculations, to fully account for costs, and to identify and consider risk. See CIO Council Best Practices Committee, *Value Measuring Methodology: How-To-Guide* (Washington, D.C.: October 2002).

them identify opportunities to achieve efficiencies and reduce duplication or migrate areas to a shared service. In addition, 1 agency encouraged OMB to continue to provide guidance and issue directives related to CIO authority and empowerment. With respect to the agencies' use of an IT valuation model, 8 agencies generally recognized the value of using such a model; however, they identified challenges in determining the appropriate model and the need to continue to refine processes and analyze the supporting cost data. Two agencies also stated that OMB should assist in facilitating and sharing IT valuation model best practices and other benchmarks among federal agencies. More specifically, 1 agency stated that OMB should assist in the development of a federal IT valuation model, and another agency suggested that best practices regarding IT valuation models should include those from private sector institutions.

As part of the 2013 OMB memorandum on PortfolioStat, OMB generally identified the same broad themes from the lessons learned documents that agencies reported. OMB has also established a page related to the 2013 PortfolioStat implementation.

Opportunities and Estimated Cost Savings Are Underreported

In separate guidance supporting the PortfolioStat initiative, OMB asked agencies to report planned cost savings and avoidance[35] associated with their consolidation and shared service initiatives through fiscal year 2015. While agencies included consolidation efforts for which they had cost savings numbers, six agencies also reported planned migration or consolidation efforts for which they had incomplete information on cost savings and avoidance.

According to OMB, agencies reported a total of 98 consolidation opportunities and $2.53 billion in planned cost savings and avoidance for fiscal years 2013 through 2015. However, OMB's overall estimate of the number of opportunities and cost savings is underreported. Among other things, it does not include the Departments of Defense and Justice because these agencies did not report their plans in the template OMB was using to compile its overall estimate. While OMB acknowledged that the $2.53 billion in planned cost savings and avoidance was

---

[35]According to OMB Exhibit 300 guidance, cost savings represents the reduction in actual expenditures while cost avoidance represents results taken from an action in the immediate time frame that will decrease costs in the future.

underreported when it issued the estimate, it did not qualify the figure quoted. Identifying any limitations or qualifications to reported figures is important in order to provide a more complete understanding of the information presented. Until OMB discloses any limitations or qualifications to the data it reports on the agency's consolidation efforts and associated savings and avoidance, the public and other stakeholders may lack crucial information needed to understand the current status of PortfolioStat and agency progress in meeting the goals of the initiative.

Our analysis of data collected from the 26 agencies shows that they are reporting 204 opportunities and at least $5.8 billion in savings through fiscal year 2015, at least $3.3 billion more than the number initially reported by OMB. See table 3 for an overview of the number of opportunities and reported cost savings and avoidance by agency. See appendix III for a detailed list of opportunities and associated savings by agency.[36]

---

[36]Subsequent to its report of agencies' planned consolidation initiatives and associated savings, OMB issued the 2013 PortfolioStat memo requiring agencies to submit an updated list of consolidation initiatives and estimated cost savings and avoidance by May 15, 2013. Some agencies reported to us that the information provided in this submission has changed from what they reported in 2012.

Table 3: Agency-Reported PortfolioStat Initiatives and Estimated Cost Savings and Avoidance

Agency	Number of initiatives	FY13 cost savings and avoidance	FY14 cost savings and avoidance	FY15 cost savings and avoidance	Total cost savings and avoidance FY13–FY15
					Dollars in millions
Agriculture	3	72.84	86.61	109.54	268.99
Commerce	8	8.19	16.60	43.01	67.80
Defense[a]	26	-	-	-	3200.00–5,300.00
DHS	15	369.27	501.25	501.25	1,371.77
Education	18	2.02	1.63	1.63	5.28
Energy	16	8.21	10.36	12.15	30.72
EPA	2	0.00	0.00	0.00	0.00
GSA	3	2.12	3.29	7.79	13.20
HHS	2	0.00	0.00	0.00	0.00
HUD	4	2.87	2.87	2.86	8.60
Interior	6	20.12	22.95	18.82	61.89
Justice[b]	12	-	-	-	35.00
Labor	12	1.75	8.59–9.89	11.96–13.26	22.30–24.90
NARA	3	-0.02	-0.02	6.47	6.43
NASA	6	19.02	23.29	24.97	67.29
NRC	8	0.00	9.60	0.00	9.60
NSF	2	0.27	0.24	0.24	0.75
OPM	2	0.00	3.50	0.00	3.50
SBA	6	0.43	0.37	0.00	0.79
SSA	7	7.23	150.15	8.78	166.16
State	4	0.00	20.00	6.00	26.00
Transportation	24	14.40	36.57	0.00	50.98
Treasury	2	24.03	67.75	101.00	192.78
USACE	2	0.50	0.50	0.50	1.50
USAID	4	3.44	5.24	8.76	17.44
VA	7	54.18	52.51	89.22	195.92
Total[c]	204	610.87	1023.84–1025.14	954.96–956.26	5824.66–7927.26

Source: GAO analysis of agency data.

Note: DHS—Department of Homeland Security; EPA—Environmental Protection Agency; GSA—General Services Administration; HHS—Department of Health and Human Services; HUD—Department of Housing and Urban Development; NARA—National Archives and Records Administration; NASA—National Aeronautics and Space Administration; NRC— U.S. Nuclear Regulatory Commission; NSF—National Science Foundation; OPM—Office of Personnel Management; SBA—Small Business Administration; SSA—Social Security Administration; USACE—U.S. Army Corps of Engineers; USAID—U.S. Agency for International Development; VA—Department of Veterans Affairs.

GAO-14-65 OMB PortfolioStat Initiative

Note: FY—fiscal year

[a]Defense did not provide information on the number of opportunities or potential cost savings to OMB in the cost savings template. This information was obtained from the department's final action plan and other referenced documentation. The department reported savings for fiscal year 2012-2015 but did not break down cost savings by initiative.

[b]Justice did not provide information on the number of opportunities or potential cost savings to OMB in the cost savings template. This information was obtained from the department's final action plan. According to officials, the $35 million total represents a rough-order-of-magnitude estimate.

[c]Numbers may not add up due to rounding.

Selected Departments' Plans Identified Billions in Potential Cost Savings Using Various Processes, but Support for These Savings Is Uneven

In their portfolio improvement plans, the five agencies selected for our review—the Departments of Agriculture, Defense, the Interior, the Treasury, and Veterans Affairs—identified a total of 52 initiatives expected to achieve at least $3.7 billion in potential cost savings or avoidance through fiscal year 2015, as well as several improvements of processes for managing their IT portfolios. To identify these opportunities, the agencies used several processes and tools, including, to varying degrees, their EA and valuation model, as recommended by OMB in its PortfolioStat guidance. More consistently using the processes recommended by OMB could assist agencies in identifying further opportunities for consolidation and shared services. In addition, four agencies did not always provide support for their estimated savings or show how it linked to the estimates. Better support for the estimated savings would increase the likelihood that these savings will be achieved.

Department of Agriculture

The Department of Agriculture (Agriculture) identified two contract consolidations—the Cellular Phone Contract Consolidation and the Enterprise Contracts for Standardized Security Products and Services—as the commodity IT investments it planned to consolidate by December 2012. In addition to these two efforts, the department identified three efforts that it reported to OMB would yield cost savings or avoidance between fiscal years 2013 and 2015 (IT Infrastructure Consolidation/Enterprise Data Center Consolidation, Enterprise IT Systems: Tier 1 Helpdesk Consolidation, and Enterprise IT Systems: Geospatial Consolidation Initiative). In addition, Agriculture identified several other opportunities for which it had yet to identify associated cost savings or avoidance. According to officials from the Office of the CIO, several of the consolidation opportunities were identified prior to the PortfolioStat initiative being launched, as part of the Secretary's initiative to streamline administrative processes. The department also identified several process improvement efforts which, while not all specific to commodity IT, would help better manage these types of investments.

Examples of the process improvement efforts include (1) recommitting to internal TechStats as a tool for evaluating all IT investments, (2) acquiring a portfolio management tool, and (3) implementing a department-wide portfolio management program that reviews major and non-major investments on a continual basis.

Agriculture officials stated that they used their EA process to identify consolidation and shared service opportunities, and that the department checks for architectural compliance throughout its governance process. For example, Agriculture's Executive IT Investment Review Board is to ensure that the department integrates information systems investment decisions with its EA and that the department's decisions comply with EA. In addition, Agriculture's Information Priorities and Investment Council is responsible for reviews of architectural compliance and for using the EA as a framework for investment decision making. These officials also stated that while the department determines the value of its IT investments through evaluation, analyses, prioritization, and scoring, it does not have a formal, documented valuation model for doing so. Having such a model would enhance the department's ability to identify additional opportunities to consolidate or eliminate low-value, duplicative, or wasteful investments.

The department also uses other processes to help manage its IT investments. For example, Agriculture has an Executive IT Investment Review Board which is to use a scoring process in ensuring the alignment of investments with strategic goals and objectives. Further, the department noted the establishment of several governance boards, and processes, such as the EA, IT acquisition approval request, and capital planning and investment control, to ensure such alignment.

Agriculture anticipates that its efforts will generate about $221 million in cost savings or avoidance for fiscal years 2012 through 2015 and provided varying degrees of support for these estimates. Specifically, for two of the four initiatives for which we requested support[37] (Cellular Phone Contract Consolidation and the IT Infrastructure Consolidation/Enterprise Data Center Consolidation), it provided support for calculations for cost savings and avoidance estimates. However,

---

[37] For each of the five agencies, we requested the supporting documentation for four efforts: the two efforts that were to be migrated by December 2012 and two efforts with the highest anticipated cost savings or avoidance for fiscal year 2013 to 2015.

these estimates did not match those provided to OMB for the 2012 PortfolioStat process. For the third initiative, Geospatial Consolidation, Agriculture did not provide support for the estimate reported to OMB as part of the 2012 PortfolioStat process; however, it noted that this current estimate is $58.76 million less than originally reported to OMB. For the fourth, a department official from the office of the Chief Information Officer said no savings were being anticipated. Documentation received from the department noted that this effort was not a cost savings initiative but a way to meet several programmatic needs: to streamline the work required for agencies procuring security services, to improve the quality and repeatability of the security products across the agencies, and to establish a process flow that ensured the department security were included in any delivered products. An Agriculture official noted challenges with calculating cost savings or avoidance but did not identify any plans to improve its cost estimating processes. A lack of support for its current estimates may make it difficult for Agriculture to realize these savings and for OMB and other stakeholders to accurately gauge its performance.

Department of Defense

The Department of Defense (Defense) identified its Unclassified Information Sharing Service/All Partner Network and the General Fund Enterprise Business System as the two commodity opportunities that would be consolidated by December 2012. In addition to these 2 efforts, Defense identified 24 other efforts that would be undertaken from 2012 to 2015 to consolidate commodity IT services. These consolidation efforts were mostly in the areas of Enterprise IT and IT infrastructure, though the department also identified a significant effort to move its major components to enterprise-wide business systems. In addition, Defense also identified several process improvements, including restructuring its IT governance boards, establishing a department IT Commodity Council, and optimizing IT services purchasing. Defense began its effort to consolidate and improve IT services in 2010 at the request of the Secretary, prior to the launch of PortfolioStat. The Defense CIO developed a 10-Point Plan for IT Modernization focused on consolidating infrastructure and streamlining processes in several commodity IT areas, including consolidating enterprise networks, delivering a department cloud environment, standardizing IT platforms, and taking an enterprise approach for procurement of common IT hardware and software. Each of the component CIOs, in coordination with the Defense CIO, was tasked with developing plans to achieve these efforts within their own component.

As part of this process, Defense utilized its EA and valuation model to determine the list of IT improvements because, according to officials from the Office of the CIO, these processes were incorporated into its existing requirements, acquisition, and planning, programming, budget, and execution processes. In particular, Defense has taken a federated approach for developing and managing its EA that is based on enterprise-level guidance, capability areas, and component architectures and is currently in the process of drafting a new EA program management plan for improvement effectiveness and interoperability across missions and infrastructure. In addition, according to a Defense official, the department has done extensive work related to implementing a valuation model, and its value engineering process for IT investments has been integrated into the department's acquisition process. Defense also has a department website devoted to providing guidance on its valuation model. Using the EA and valuation model increases the likelihood that the department will identify a comprehensive list of opportunities for consolidation.[38]

Defense's CIO estimates that the consolidation efforts will save between $3.2 billion and $5.2 billion through fiscal year 2015, and result in efficiencies between $1.3 billion and $2.2 billion per year beginning in fiscal year 2016. Defense provided its most recent estimates for the four initiatives for which we requested support (Unclassified Information Sharing Service/All Partner Access Network, data center consolidation, enterprise software purchasing, and General Fund Enterprise Business System) but was unable to show how these estimates were calculated. For the first initiative, the issue paper showing the calculations of estimated savings was reportedly classified and we therefore decided not to obtain a copy.[39] For the other three initiatives, an official from the Office of the CIO stated that there was not support available at the department level. Each component reportedly used its existing planning, programming, budget and execution process, and associated systems to

---

[38]Since 1995, we have designated the Department of Defense's business systems modernization program as high risk because of its vulnerability to fraud, waste, abuse, and mismanagement, and because of missed opportunities to achieve greater efficiencies. Since then, we have made a series of recommendations aimed at strengthening its institutional approach to modernization and reducing the risk associated with key investments. See most recently GAO, *DOD Business Systems Modernization: Further Actions Needed to Address Challenges and Improve Accountability*, GAO-13-557 (Washington, D.C.: May 17, 2013).

[39]We were informed the document was classified at the end of our review. Due to the timing, we decided not to obtain a copy or select another initiative for review.

determine a overall budget and then identified estimated cost savings or avoidance related to the commodity initiatives, which were then aggregated by the department. The official also reported that, because the department's accounting systems do not collect information at the level of granularity required for reporting on the PortfolioStat initiative (e.g., by commodity IT type), it is difficult to show how numbers were calculated or how they changed over time. In addition, because component-level systems do not collect commodity IT data, it had generally been a challenge for the department to determine cost savings for commodity IT as OMB required. While we recognize the challenges the department faces in obtaining the support for consolidation opportunities identified by its components, obtaining it is critical to ensuring that planned savings and cost avoidance are realized. This is important considering the size of Defense's projected savings.

Department of the Interior

The Department of the Interior (Interior) identified two commodity IT investments in its action plan and other supporting documentation—Financial and Business Management System (Deployments 7&8) and Enterprise Forms System—that it planned to consolidate by December 2012. For fiscal years 2013 to 2015, Interior identified four additional consolidation opportunities—cloud e-mail and collaboration services, enterprise eArchive system, circuit consolidation, and the Networx telecommunications contract. Interior also identified its "IT Transformation" initiative as a source of additional savings beyond 2015. This initiative is one of the management priorities which, according to officials, Interior has been focusing on to drive efficiency, reduce costs, and improve services. It is intended to streamline processes within the department, to include a single e-mail system for the department, telecommunications, hosting services, and an enterprise service desk (help desk).

Interior has also identified efforts to improve processes for managing its portfolio. Specifically, it is working to fully implement its EA and to align the IT portfolio more closely with the department's business priorities and performance goals. In addition, in fiscal year 2010, Interior centralized authority for the agency's IT—which had previously been delegated to its offices and bureaus—under the CIO. This consolidation gave the CIO responsibilities for improving the operating efficiencies of the organizational sub-components and Interior as a whole. Interior is also establishing several new IT Investment governance boards to make recommendations to the CIO for review and approval.

To identify its consolidation opportunities, Interior officials from the Office of the CIO stated they used their EA. Specifically, the department established an EA team and a performance-driven prioritization framework to measure its IT Transformation efforts. The EA team takes a "ruthless prioritization" approach to align the department's priorities with the IT Transformation goals. The priorities are evaluated by IT Transformation goals and expected outcomes, and supported by successive versions of architectures, plans, and solutions. In addition to using the EA, officials from the Office of the CIO stated that the department leveraged a set of investment processes to identify wasteful, duplicative, and low-value investments, which includes the use of road maps it has developed for different functional areas. Collectively, these processes are integrated into the department's capital planning and investment control process in order to ensure that the portfolio of IT investments delivers the desired value to the organization.

Interior officials from the Office of the CIO also reported using its IT investment valuation process which it has been maturing while also balancing changes to its IT governance process. More specifically, the department uses the Value Measuring Methodology, recommended by the federal CIO Council, to score its bureaus' budget requests. Based on these assessments, a risk-adjusted value score is assigned to each major investment. These scores are used to identify funding levels across Interior's IT portfolio, with risk being viewed from the standpoint of the "probability of success" for the investment. By making use of the EA and investment valuation process as recommended by OMB, Interior has enhanced its ability to identify opportunities to consolidate or eliminate duplicative, low-value, and wasteful investments.

Interior anticipates its PortfolioStat efforts will generate approximately $61.9 million in savings and cost avoidance through fiscal year 2015 and provided adequate support for these estimates. Specifically, for the Financial and Business Management System, Interior provided calculations for the savings for each year from fiscal year 2012 to fiscal year 2016. For the other three initiatives—Electronic Forms System, Networx Telecommunications, and Cloud E-mail and Collaboration Services—Interior provided estimated savings for fiscal year 2013, the first year in which savings are anticipated, which were based on the difference between the fiscal year 2012 baseline and lower costs that had been achieved through the department's strategic sourcing initiative, and explained that these savings were expected to be realized each year after—through fiscal year 2015. Having well-supported estimates increases the likelihood that Interior will realize its planned savings and provides OMB and other stakeholders with greater visibility into the department's performance.

Department of the Treasury

The Department of the Treasury (Treasury) identified two new shared service opportunities—the Invoice Processing Platform and the DoNotPay Business Center—as the two commodity IT investments it planned to consolidate by December 2012; Treasury also reported to OMB that these efforts would yield cost savings and avoidance for fiscal years 2013 through 2015. In addition, Treasury identified six consolidation opportunities it anticipated would generate savings between fiscal years 2012 and 2014, and two others which did not have associated cost savings. These consolidation opportunities include those classified as Business Systems, IT Infrastructure, and Enterprise IT. Treasury also described several process improvement efforts which, while not specific to commodity IT, will help better manage these types of investments. Examples of the process improvement efforts include establishing criteria for robust reviews of investments, refining the department's valuation and risk models and incorporating these models into the business case template at Treasury's departmental offices, and launching an IT cost model working group to refine Treasury's IT cost model. Treasury has also proposed additional steps in its Departmental Offices' IT governance process and investment life cycle to evaluate the alignment of investments with its strategic goals and objectives. With respect to EA, in July 2013, Treasury established a Treasury Departmental Technical Advisory Working Group. According to its charter, the working group will be responsible for, among other things, ensuring the alignment of programs and projects with Treasury's existing technologies or EA. More specifically, all new and existing investments are expected to be reviewed and approved by the working group to ensure such compliance.

Treasury officials from the Office of the CIO stated they had not used the EA or a valuation model to identify their consolidation opportunities. In addition, Treasury has yet to develop a valuation model for assessing the value of its IT investments. According to officials, Treasury's efforts to develop a valuation model are 30 to 40 percent complete. Further, while it has efforts underway within its Departmental Offices to develop models for assessing value, cost, and risk, Treasury has not documented its value engineering process and associated models. According to the officials, the department's consolidation opportunities were identified through innovative ideas from the bureaus that were driven by current budget constraints. While the identification of these opportunities is not centrally managed or controlled, Treasury reported that it is currently developing a systematic process for promoting innovative ideas from its bureaus.

According to Treasury, it uses other processes to help manage IT investments, including a process for evaluating the alignment of investments with its strategic goals and objectives via its investment review boards at both the department-wide and departmental office levels. Further, Treasury has noted that investments' alignment with the mission is considered during the annual planning cycle (for existing and new investments), and during individual investment/project reviews (for every major investment). While Treasury identified consolidation and shared service opportunities through innovative ideas from its bureaus, utilizing the EA and valuation model could assist Treasury in identifying additional opportunities for cost savings.

Treasury anticipates it will generate $56.49 million in savings from fiscal years 2012 through 2014 and provided varying degrees of support for these estimates.[40] Specifically, for two of the three initiatives that we reviewed supporting documentation for,[41] one initiative (DoNotPay Business Center) had supporting assumptions and calculations; however, these calculations support earlier estimates Treasury reported for this initiative, and not its more recent estimates. Treasury did not provide documentation to support the cost estimates for the two remaining efforts (Fiscal IT Data Center Consolidation and Business Process Management Status). Without support for its estimates, Treasury may be challenged in realizing planned savings, and OMB and other stakeholders will be hindered in evaluating its progress.

Department of Veterans Affairs

The Department of Veterans Affairs (VA) identified its VA Server Virtualization and Elimination of Dedicated Fax Servers as the two commodity IT investments it planned to consolidate by December 2012. In its PortfolioStat submission to OMB, VA identified five additional

---

[40]These estimates are based on Treasury's May 2013 integrated data collection to OMB and do not include cost savings or avoidance of $116. 25 million which Treasury anticipates will be achieved by other federal agencies using its Invoice Processing Platform.

[41]We did not review supporting documentation for the cost savings/avoidance estimates associated with the Invoice Processing Platform due to the fact that these cost savings/avoidance will be achieved by other agencies, not Treasury.

consolidation opportunities it anticipated would generate savings between fiscal years 2013 and 2015 (enterprise license agreement, standardization of spend planning and consolidation of contract, voice over internet protocol, vista data feeds, and one CPU policy). VA also described several process improvement efforts in its action plan that, while not specific to commodity IT, are intended to help better manage these types of investments. These improvement efforts include updating its EA process and establishing a Project Management Accountability System that supports project planning and management control and responsibilities for IT investments.

VA officials from the Office of the CIO stated that they did not use their EA (which the department is still maturing) or their valuation model to identify their consolidation opportunities. Instead, they stated that VA uses its Ruthless Reduction Taskforce as the main mechanism for identifying IT commodity consolidation opportunities. The task force's function is to ensure redundant functionality is reduced or eliminated and to recommend the reallocation of funds from low-value projects to higher priorities. Through its operational analysis process, it looks for excessive expenditures to determine whether there are redundancies and therefore opportunities to consolidate into a larger contract or service.[42]

While the task force is the main mechanism used to identify consolidation opportunities, VA officials from the Office of the CIO stated that the department uses other OMB-recommended processes to help it identify and prioritize other IT investments. For example, VA has documented processes for evaluating the alignment of investments with its strategic goals and objectives via its multiyear planning process and its senior investment review boards. Specifically, the department's multiyear planning process provides a framework for identifying near- and long-term priorities and opportunities for divestiture, reduction, re-investments, and

---

[42]The Ruthless Reduction Task Force's approach includes: (1) utilizing subject matter expert opinions to examine analytical and engineering processes; perform an analysis of alternatives through business case analysis, financial analysis, and cost analysis; and re-examine/modify approaches to achieving cost savings/avoidance; (2) reviewing the use of IT equipment, software, and systems to determine value, gauge efficiency, and formulate recommendations for potential cost savings or cost avoidance; and, (3) establishing a VA-wide process to continuously solicit and collect ideas and recommendations to improve the efficiency of IT projects, programs, and the acquisition and allocation of equipment; and assisting project managers in the development of VA-wide cost savings measures for presentation to the Assistant Secretary.

expansion of IT priorities and capabilities and timetables. To support this and other planning processes, VA has established several IT Investment governance boards that are intended to provide a framework for investment decision making and accountability to ensure IT initiatives meet the department's strategic and business objectives in an effective manner. While VA has identified many opportunities to consolidate commodity IT investments and move to shared services through its Ruthless Reduction Task Force and other processes, making use of its EA and valuation model could help identify additional opportunities.

VA estimates that the consolidation opportunities it reported to OMB will generate about $196 million in savings from fiscal years 2013 through 2015. However, we could not verify the support for some of the estimates. In particular, for two of the four initiatives for which we requested support (Server Virtualization and Eliminate Dedicated Fax Servers Consolidation), VA provided support for calculations for cost savings and avoidance estimates. However, these estimates did not match those provided to OMB for the 2012 PortfolioStat process. For the third initiative, Renegotiate Microsoft Enterprise License Agreement, VA did not provide detailed support but instead provided a written explanation for an overall cost avoidance figure of $161 million that was agreed to by VA's Deputy Chief Information Officer for Architecture, Strategy and Design and VA's Deputy Assistant Secretary for Information Technology Management and Chief Financial Officer for the Office of Information Technology. For the fourth initiative (one CPU policy), VA stated that the initiative was no longer a stand-alone project but had been subsumed by the Field Office Mobile Workers and Telework Support Agreement and that the economic justification for this consolidation effort had not yet been completed.

Officials reported that in general the lack of a strong cost estimation process is the main challenge the department faced in estimating cost savings, even though VA's Ruthless Reduction Task Force does have a process in place for performing cost estimates for the initiatives that the task force reviews. VA officials stated that they plan to address improving their IT cost estimation process issue with VA's executive leadership team, but did not provide a time frame for doing so. For the near term, VA recently hired an operations research analyst to assist IT staff who lack experience with cost and savings estimation activities and plans to hire two more analysts. Without support for its estimates, VA will have less assurance that it can realize planned cost savings and avoidance, and OMB and stakeholders will be hindered in evaluating its progress.

OMB's Plans Outline PortfolioStat Improvements but Do Not Address All Issues with Agencies' Efforts

OMB has outlined several planned improvements to the PortfolioStat process in a memo issued in March 2013[43] that should help strengthen federal IT portfolio management and address key issues we have identified with agencies' efforts to implement the initiative. In particular, OMB has changed its reporting requirements, requiring agencies to report on progress made on a quarterly basis. In addition, agencies will also be held accountable for their portfolio management as part of annual PortfolioStat sessions. However, selective OMB efforts could be strengthened to improve the PortfolioStat process and ensure agencies achieve identified cost savings, including addressing issues related to existing CIO authority at federal agencies, and publically reporting on agency-provided data.

OMB's plans identify a number of improvements that should help strengthen IT portfolio management and address key issues we have identified:

Agency reporting on PortfolioStat progress: OMB's memorandum has consolidated previously collected IT plans, reports, and data calls into three primary collection channels—an information resources management strategic plan,[44] an enterprise road map,[45] and an integrated data collection channel.[46] As part of this reporting requirement, agencies will be required to provide updates on their progress in meeting key OMB

---

[43]OMB, *Memorandum for the Heads of Executive Departments and Agencies: Fiscal Year 2013 PortfolioStat Guidance: Strengthening Federal IT Portfolio Management*, M-13-09 (Washington, D.C.: Mar. 27, 2013).

[44]OMB, *Management of Federal Information Resources*, Circular A-130 (Washington, D.C.: Nov. 28, 2000). According to OMB Circular A-130, an agency's information resources management strategic plan should describe how information resources management activities help accomplish agency missions, and ensure that information resource management decisions are integrated with organizational planning, budget, procurement, financial management, human resources management, and program decisions.

[45]OMB, *Increasing Shared Approaches to Information Technology Services* (Washington, D.C.: May 2, 2012). The enterprise road map is to include a business and technology architecture, an IT asset inventory, a commodity IT consolidation plan, a line of business service plan, and an IT shared service plan.

[46]The integrated data collection channel will be used by agencies to report structured information, such as progress in meeting IT strategic goals, objectives, and metrics, as well as cost savings and avoidances resulting from IT management actions.

requirements related to portfolio management best practices, which address issues identified in this report.

- Agencies must describe how their investment review boards coordinate between investment decisions, portfolio management, EA, procurement, and software development methodologies to ensure that IT solutions meet business requirements, as well as identify areas of waste and duplication wherever consolidation is possible.

- Agencies are to describe the valuation methodology used in their governance process to comparatively evaluate investments, including what criteria and areas are assessed, to ensure greater consistency and rigor in the process of selecting, controlling, and evaluating investments an agency decides to fund, de-fund, or terminate.

- Agencies must report their actual and planned cost savings and avoidances, as well as other metrics, achieved or expected through the implementation of efforts such as agency migration to shared services and cloud solutions, the consolidation of commodity IT, and savings achieved through data center consolidation. In addition, agencies are to describe their plans to re-invest savings resulting from consolidations of commodity IT resources (including data centers).

In addition, agencies will now be required to report the status of their progress in implementing PortfolioStat on a quarterly basis. Agency integrated data collections were first required to be submitted in May 2013 and will be updated quarterly beginning in August 2013, with subsequent updates on the last day of November, and February of each fiscal year. Requiring agencies to provide consolidated reports on their progress in meeting key initiatives should help OMB to better manage these initiatives.

Holding agencies accountable for portfolio management in PortfolioStat sessions: Moving forward, the PortfolioStat sessions held with agency stakeholders and OMB officials are intended to involve discussions of agency efforts related to several ongoing initiatives and their plans to implement key OMB guidance, such as guidance on CIO authorities, in order to help agencies mature their management of IT resources. Specifically, OMB plans to use the documentation and data submitted by the agencies in May 2013 to determine the state of each agency's IT portfolio management, such as the use of an EA and valuation methodology, and develop areas OMB identifies as the most appropriate opportunities for agencies to innovate, optimize, and protect systems and data. Based on the session, OMB and the agency are expected to identify and agree on actionable next steps and specific time frames for the actions to be taken, which OMB intends to formalize and transmit in a memorandum to the agency within 2 weeks of the completed session, and no later than August 31, 2013. Upon receipt of the action item memorandum, agency PortfolioStat leads are to work with OMB to establish follow-up discussions as appropriate to track progress against action items identified. Deviation from the committed schedule will trigger a requirement for follow-up briefings by the agency to the Federal CIO no less frequently than quarterly, until corrective actions have been implemented or the action item is back on schedule.

OMB's efforts to follow up with agencies on a regular basis are critical to ensuring the success of these efforts. We have previously reported that OMB-led TechStat sessions have enabled the government to improve or terminate IT investments that are experiencing performance problems by focusing management attention on troubled projects and establishing clear action items to turn the projects around or terminate them.[47] By having similar sessions focusing on agency IT portfolios, OMB can hold agencies accountable for their ongoing initiatives to consolidate or eliminate duplicative investments and achieve significant cost savings.

Improving analytical capabilities: OMB expects to collect information from agencies as part of PortfolioStat and use a variety of analytical resources to evaluate the data provided, track agency progress each quarter, and determine whether there are any areas for improvement to the process. In addition, OMB plans to provide this information to Congress as part of the quarterly report it is required to submit to the Senate and House Appropriations Committees on savings achieved by OMB's government-wide IT reform efforts. Analyzing and reporting data on agencies' efforts to implement the PortfolioStat initiative will help OMB to provide more oversight of these efforts and hold agencies accountable for information reported in the quarterly reports.

Although OMB's planned improvements should help strengthen the PortfolioStat initiative going forward, they do not address some of the shortcomings with efforts to implement the initiative identified in this report:

---

[47] GAO-13-524.

Addressing issues with CIO authority: While OMB's memorandum has indicated that agencies must now report on how their policies, procedures, and CIO authorities are consistent with OMB Memorandum 11-29, "Chief Information Officer Authorities,"[48] as noted earlier, OMB's prior guidance and reporting requirements have not been sufficient to address the implementation of CIO authority at all agencies. In addition, OMB's 2013 PortfolioStat guidance does not establish deadlines or metrics for agencies to demonstrate the extent to which CIOs are exercising the authorities and responsibilities provided by the Clinger-Cohen Act and OMB guidance, which, as we have previously recommended, are needed to ensure accountability for acting on this issue,[49] nor does it require them to disclose any limitations CIOs might have in their ability to exercise their authority. Until CIOs are able to exercise their full authority, they will be limited in their ability to implement PortfolioStat and other initiatives to improve IT management.

Reporting on action plan items that were not addressed: In OMB's 2013 memorandum, agencies are no longer required to submit separate commodity IT consolidation plans as in 2012 but are to identify the progress made in implementing portfolio improvements as part of the broader agency reporting requirement mentioned above. While OMB's shift to requiring agencies to report on progress now is reasonable given the goals of PortfolioStat, it was based on the assumption that agencies would develop robust action plans as a foundation last year. However, as noted earlier, the submitted agency final action plans were incomplete in that they did not always address all the required elements. Going forward, it will be important for agencies to address the plan items required. In addition, until OMB requires agencies to report on the status of these items, it may not have assurance that these agencies' plans for making portfolio improvements fully realize the benefits of the PortfolioStat initiative.

Ensuring agencies' commodity IT baselines are complete, and reporting on the status of 2012 migration efforts: OMB's 2013 guidance does not require agencies to document how they verified their

---

[48]OMB M-11-29 states that CIOs must be empowered by the agency head to have authority over IT governance, commodity IT systems, information security, and IT program management in order to drive efficiencies.

[49]GAO-11-634.

commodity IT baseline data or disclose any limitations of these data or to report on the completion of their two 2012 migration efforts. Without such requirements, it will be more difficult for OMB to hold agencies accountable for identifying and achieving potential cost savings.

Publically reporting agency PortfolioStat data: Finally, we have previously reported that the public display of agencies' data allows OMB, other oversight bodies, and the general public to hold the agencies accountable for results and progress.[50] While OMB officials have stated that they intend to make agency-reported data and the best practices identified for the PortfolioStat effort publicly available, they have not yet decided specifically which information they will report. Until OMB publicly reports data agencies submit on their commodity IT consolidation efforts, including planned and actual cost savings, it will be more difficult for stakeholders, including Congress and the public, to monitor agencies' progress and hold them accountable for reducing duplication and achieving cost savings.

Conclusions

OMB's PortfolioStat initiative offers opportunities to save billions of dollars and improve the way in which agencies manage their portfolios. While agencies implemented key PortfolioStat requirements, including establishing a commodity IT baseline and documenting a final action plan to consolidate commodity IT, shortcomings in their implementation of these requirements could undermine the savings the PortfolioStat effort is expected to achieve. First, reported limitations in CIOs exercising authority over the IT portfolios at six of the agencies suggests that more needs to be done to empower CIOs to improve management and oversight of agency IT resources. Second, not including all IT investments in their EA or developing complete commodity IT baselines limits agencies' ability to identify further opportunities for reducing wasteful, duplicative, or low-value investments. Third, not addressing key elements in action plans for implementing the PortfolioStat initiative increases the likelihood that agencies will not achieve all the intended benefits. Finally, following through on commitments to migrate or consolidate investments is critical to ensuring accountability for results. Regarding estimated savings and cost avoidance, the significant understatement—by at least

---

[50]GAO, *Information Technology: OMB's Dashboard Has Increased Transparency and Oversight, but Improvements Needed,* GAO-10-701 (Washington, D.C.: July 2010).

$2.8 billion—of OMB's reported figures highlights the importance of ensuring the accuracy of data and disclosing any limitations or qualifications on reported savings.

The identification by five agencies—the Departments of Agriculture, Defense, the Interior, the Treasury, and Veterans Affairs—of 52 initiatives and more than $3.7 billion in potential cost savings or avoidance through fiscal year 2015 demonstrates the significant potential of portfolio improvements to yield ongoing benefits. Making greater use of their EA and valuation model to identify consolidation opportunities, as recommended by OMB, could assist agencies in identifying additional opportunities. In addition, better support for the estimates of cost savings associated with the opportunities identified would increase the likelihood that these savings will be achieved.

OMB's planned improvements to the PortfolioStat process outlined in its March 2013 guidance—such as streamlining agency reporting on progress in implementing the process and holding agencies accountable for these efforts in PortfolioStat sessions—should help the office provide better oversight and management of government-wide efforts to consolidate commodity IT. However, OMB's plans do not address key issues identified in this report, which could strengthen the PortfolioStat process. In particular, addressing issues of CIO authority by working directly with agency leadership to establish time lines and metrics for implementing existing guidance, requiring agencies to report on the reliability of their commodity baseline data and the progress of all their consolidation efforts, and making data on agencies' progress in consolidating commodity IT and achieving cost savings publicly available will be essential to PortfolioStat's success in reducing duplication and maximizing the return on investment in federal IT.

Recommendations for Executive Action

To help ensure the success of PortfolioStat, we are making six recommendations to OMB. We recommend that the Director of the Office of Management and Budget and the Federal Chief Information Officer require agencies to fully disclose limitations their CIOs might have in exercising the authorities and responsibilities provided by law and OMB's guidance. Particular attention should be paid to the Departments of Health and Human Services, and State; the National Aeronautics and Space Administration; the Office of Personnel Management; and the U.S. Agency for International Development, which reported specific limitations with the CIO's authority.

In addition, we recommend that the Director of the Office of Management and Budget direct the Federal Chief Information Officer to

- require that agencies (1) state what actions have been taken to ensure the completeness of their commodity IT baseline information and (2) identify any limitation with this information as part of integrated data collection quarterly reporting;

- require agencies to report on the progress of their two consolidation efforts that were to be completed by December 2012 as part of the integrated data collection quarterly reporting;

- disclose the limitations of any data reported (or disclose the parameters and assumptions of these data) on the agencies' consolidation efforts and associated savings and cost avoidance;

- require that agencies report on efforts to address action plan items as part of future PortfolioStat reporting; and

- Improve transparency of and accountability for PortfolioStat by publicly disclosing planned and actual data consolidation efforts and related cost savings by agency.

We are also making 58 recommendations to 24 of the 26 departments and agencies in our review to improve their implementation of PortfolioStat requirements. Appendix IV contains these recommendations.

Agency Comments and Our Evaluation

We provided a draft of this report to OMB and the 26 executive agencies in our review for comment and received responses from all 27. Of the 27, 12 agreed with our recommendations directed to them, 5 disagreed or partially disagreed with our recommendations directed to them, 4 provided additional clarifying information, and 6 (the Departments of Education, Labor, Transportation, and Treasury; the Small Business Administration; and the U.S. Agency for International Development) stated that they had no comments.[51] Several agencies also provided technical comments, which we incorporated as appropriate. The agencies' comments and our responses are summarized below.

---

[51]We did not make any recommendations to the Department of Education.

- In e-mail comments from the Federal Chief Information Officer, OMB generally agreed with three of our recommendations and disagreed with three. Specifically, OMB agreed with the recommendation to require agencies to disclose limitations their CIOs might have in exercising the authorities and responsibilities provided by law and OMB guidance but stated that it had already addressed this issue as part of its fiscal year 2013 PortfolioStat process. Specifically, according to OMB, its fiscal year 2013 PortfolioStat guidance[52] required agencies to describe how their policies, procedures, and authorities implement CIO authorities, consistent with OMB Memorandum 11-29, as part of either the information resources management plan or enterprise roadmap they were instructed to submit. OMB stated that it reviewed and analyzed agencies' responses and discussed limitations to CIOs' authorities directly with agencies during the PortfolioStat sessions in cases where it determined that such limitations existed. However, OMB did not provide documentation supporting its reviews or discussions with agencies. In addition, as we note in our report, requiring agencies to fully disclose limitations their CIOs may have in exercising the authorities and responsibilities provided by law and OMB guidance should provide OMB information crucial to understanding and addressing the factors that could prevent agencies from successfully implementing the PortfolioStat initiative. For these reasons, we are maintaining our recommendation.

OMB stated that it agreed with our recommendation to require that agencies (1) state what actions have been taken to ensure the completeness of their commodity IT baseline information and (2) identify any limitations with this information as part of the integrated data collection quarterly reporting. It acknowledged the value in ensuring the completeness and in understanding the limitations of agency-produced artifacts and stated it would continue to dedicate resources to validating agency savings associated with federal IT reform efforts prior to presenting these savings to Congress. OMB also stated that it would modify its analytical process to cite these limitations when producing PortfolioStat reports in the future.

OMB generally agreed with the recommendation to require agencies to report on the progress of the two consolidation efforts they were to

---

[52] OMB M-13-09.

complete by December 2012 and stated that, to the extent feasible, it would dedicate resources to analyzing this information.

OMB disagreed with our recommendation to disclose the limitations of any data reported on the agencies' consolidation efforts and associated cost savings and avoidance, stating that it had disclosed limitations on data reported and citing three instances of these efforts. While we acknowledge that OMB reported limitations of data regarding consolidation efforts in these cases, the information reported did not provide stakeholders and the public with a complete understanding of the information presented. For example, OMB did not disclose that information from the departments of Defense and Justice was not included in the consolidation estimates reported, which, considering the scope of Defense's efforts in this area (at least $3.2 billion), is a major gap. As noted in our report, OMB's disclosure of limitations of or qualifications to the data it reports would provide the public and other stakeholders with crucial information needed to understand the status of PortfolioStat and agency progress in meeting the goals of the initiative. Therefore, we stand by our recommendation.

OMB also disagreed with our recommendation to require agencies to report on efforts to address action plan elements as part of future OMB reporting, stating that it had found that 24 of 26 agencies had completed their plans. OMB further stated that it continuously follows up on the consolidation efforts identified in the plans and, where savings have been identified, reports this progress to Congress on a quarterly basis. However, our review of the 26 agency action plans found 26 instances where a required element (e.g., consolidation of commodity IT spending under the CIO) was not addressed and 26 instances where a required element was only partially addressed--an assessment with which agencies agreed. As noted in our report, addressing all the required elements would better position agencies to fully realize the intended benefits of the PortfolioStat initiative, and they should therefore be held accountable for reporting on them as required in OMB memo M-12-10. Accordingly, we stand by our recommendation.

Finally, OMB disagreed with our recommendation to improve transparency and accountability for PortfolioStat by disclosing consolidation efforts and related cost savings by agency. Specifically, OMB stated that this recommendation does not adequately account for the work it currently performs to ensure accountability for and transparency of the process through its quarterly reporting of identified savings to Congress. It further stated that some details are deliberative or procurement sensitive and it would therefore not be appropriate to disclose them. However, while OMB currently reports realized savings by agency on a quarterly basis, these savings are not measured against planned savings. Doing this would greatly enhance Congress's insight into agencies' progress and hold them accountable for reducing duplication and achieving planned cost savings and would not require reporting deliberative or procurement-sensitive information. Therefore, we stand by our recommendation.

- In written comments, the U.S. Department of Agriculture concurred with the content of our report. The department's comments are reprinted in appendix V.

- In written comments, the Department of Commerce concurred with our recommendations but disagreed with our statement that the CIO only has explicit authority over major IT investments. Commerce cited a June 21, 2012, departmental memo on IT portfolio management that it believes provides the CIO with explicit authority to review any IT investment, whether major or non-major. Our statement regarding the limitations on the CIO's authority was based on information reported by the department to OMB in May 2012 and confirmed with officials from the Commerce Office of the CIO during the course of our review. However, we agree that the June 2012 memo provides the CIO with explicit authority to review all IT investments. Accordingly, we have removed the original statement noting limitations from the report and also removed Commerce from the list of departments OMB should require to disclose CIO limitations. The Department of Commerce's comments are reprinted in appendix VI.

- In its written response, the Department of Defense provided comments for both the department and the Army Corps of Engineers. It concurred with one of the three recommendations made to Defense, partially concurred with another and disagreed with the third. Specifically, the department concurred with our recommendation to obtain support from the relevant component agencies for the estimated savings for fiscal years 2013 to 2015 for the data center consolidation, enterprise software purchasing, and General Fund Enterprise Business System initiatives. It partially concurred with our recommendation to develop a complete commodity baseline, stating that the department has efforts under way to further refine the baseline. Since these efforts have not yet been completed, we are maintaining our recommendation. The department did not concur with our recommendation to fully describe the consolidation of commodity IT spending under the CIO in future OMB reporting. The department stated that it did not intend to follow OMB's guidance to consolidate commodity IT spending under the CIO because this approach would not work within the department's federated management process. However, our recommendation was not to implement OMB's guidance, but rather to address the element in the plan as required by either describing the steps it will take to implement it or explaining why it will not or cannot implement it. DOD did neither of these and instead was silent on the subject. We therefore stand by our recommendation. The department concurred with both of the recommendations we made to the Army Corps of Engineers. The department's comments are reprinted in appendix VII.

- In written comments, the Department of Energy concurred with our recommendation to fully describe PortfolioStat action plan elements in future OMB reporting and stated that the department was committed to increasing the CIO's oversight and authority for federal commodity IT investments. The department also noted that our statement that the "department has no direct authority over IT investments in two semi-autonomous agencies (the National Nuclear Security Administration and the Energy Information Administration)" should be clarified to say that it is the department CIO who does not have this authority. We found support for this clarification in documentation we had already received and therefore made it as requested. The department's comments are reprinted in appendix VIII.

- In written comments, the Environmental Protection Agency generally agreed with two of the three recommendations we made and generally disagreed with the third. Specifically, the Environmental Protection Agency generally agreed with our recommendations to (1) fully describe three PortfolioStat action plan elements and (2) report on the agency's progress in consolidating the managed print services and strategic sourcing of end user computing to shared services as part of the OMB integrated data collection quarterly reporting until completed. The agency disagreed with our recommendation to develop a complete commodity IT baseline, stating that it had provided a complete baseline to OMB on August 31, 2012, and had also reported to us during our review that the information was current and complete at the time of submission. During our review, we found that the Environmental Protection Agency did not have a process in place to ensure the completeness of the information in the baseline. Without appropriate controls and processes in place to confirm this, the Environmental Protection Agency cannot be assured that its data are complete. We therefore stand by our recommendation. The Environmental Protection Agency's comments are reprinted in appendix IX.

- In written comments, the General Services Administration agreed with our findings and recommendations and stated it would take action as appropriate. The agency's comments are reprinted in appendix X.

- In written comments, the Department of Homeland Security disagreed with our recommendation to fully describe its efforts related to consolidating commodity IT spending under the CIO in future OMB reporting, stating that the department had already addressed this recommendation. Specifically, the department stated that it had included updated information on this topic in its fiscal year 2013 Information Resources Management Plan that was submitted to OMB in May 2013. We reviewed the Information Resources Management Plan and agree that it addresses our recommendation. We therefore removed the recommendation from the report. The department's comments are reprinted in appendix XI.

- In written comments, the Department of Housing and Urban Development concurred with our recommendations and stated it would provide more definitive information with timelines once the final report had been issued. The department's comments are reprinted in appendix XII.

- In e-mail comments, the Department of the Interior's GAO Audit Liaison stated that the department generally concurred with our findings and recommendations. However, the department recommended revising the criteria we used to assess whether agencies met the requirement to develop a commodity IT baseline (depicted in table 1) to reflect whether or not an agency had developed a baseline instead of whether that baseline was complete. The department stated that a validation was not being performed on how all agencies responded to the question and agencies that answered truthfully were being penalized for responding honestly. We recognize that agencies were not required to report on the completeness of the commodity IT baseline information they submitted to OMB; for this reason, we have recommended that OMB require agencies to state what actions have been taken to ensure the completeness of their commodity IT baseline information and identify any limitations with this information as part of the integrated data collection quarterly reporting.

- In e-mail comments, an official from the Department of Justice's Audit Liaison Group stated that all references to the department were factually correct.

- In written comments, the National Aeronautics and Space Administration concurred with our recommendations and noted the agency will take actions to address them. The agency's comments are reprinted in appendix XIII.

- In written comments, the National Archives and Records Administration concurred with our recommendation and stated that it would include updated or new descriptions of the elements of the PortfolioStat action plan in future OMB reporting. The agency's comments are reprinted in appendix XIV.

- In written comments, the National Science Foundation stated that it generally agreed with our characterization of the agency's PortfolioStat status and would update its PortfolioStat action plan as appropriate to more fully describe the two elements that we noted were not fully addressed. Regarding our recommendation to complete the consolidation of e-mail services to shared services, the agency stated that this effort was completed in August 2013. After reviewing additional documentation provided, we agree that the agency has met the requirement. We modified the report as appropriate, and removed the recommendation. The National Science Foundation's comments are reprinted in appendix XV.

- In e-mail comments, the U.S. Nuclear Regulatory Commission's GAO Audit Liaison stated that the agency was generally in agreement with our report.

- In written comments, the Office of Personnel Management concurred with our recommendations and noted that the agency will provide updated information on efforts to address them to OMB on November 30, 2013. The agency's comments are reprinted in appendix XVI.

- In written comments, the Social Security Administration agreed with one recommendation and disagreed with the other. The agency disagreed with our recommendation to develop a complete commodity IT baseline, stating that it believed its commodity baseline data to be complete and accurate. However, our review found that the Social Security Administration did not have a process in place to ensure the completeness of the information in the baseline. Without appropriate controls and processes in place to confirm the completeness of data, the Social Security Administration cannot be assured that its data are complete. The agency also acknowledged that it needed to document a process for demonstrating the completeness of its baseline data. Consequently, we stand by our recommendation. The Social Security Administration's comments are reprinted in appendix XVII.

- In written comments, the Department of State stated that it concurred with our report and would develop specific responses to each of the three recommendations we made once the report is published. However, related to our recommendation to complete the consolidation of the Foreign Affairs Network and content publishing and delivery services, the department stated that it has already consolidated more than two commodity IT areas per OMB Memorandum M-11-29. While we acknowledge that it has made efforts in this area, during our review the department changed what it considered the two commodity areas to be consolidated by December 2012 several times before stating that the two efforts were the Foreign Affairs Network and content publishing and delivery services. Based on this determination, we assessed the status of these two efforts and confirmed that neither had been completed as of August 2013. In addition, the department did not provide any documentation to support that it had consolidated more than two commodity IT areas. We therefore stand by our recommendation. The Department of State's comments are reprinted in appendix XVIII.

- In written comments, the Department of Veterans Affairs concurred with our four recommendations, stating the department is taking steps to manage its investment portfolio more effectively and has developed an action plan to address each recommendation. The department's comments are reprinted in appendix XIX.

We are sending copies of this report to interested congressional committees, the Director of the Office of Management and Budget, the secretaries and agency heads of the departments and agencies addressed in this report, and other interested parties. In addition, the report will be available at no charge on GAO's website at http://www.gao.gov.

If you or your staffs have any questions on the matters discussed in this report, please contact me at (202) 512-9286 or pownerd@gao.gov. Contact points for our Offices of Congressional Relations and Public Affairs may be found on the last page of this report. GAO staff who made major contributions to this report are listed in appendix XX.

David A. Powner
Director, Information Technology
 Management Issues

Appendix I: Objectives, Scope, and Methodology

Our objectives were to (1) determine the status of efforts to implement key required PortfolioStat actions, (2) evaluate selected agencies' plans for making portfolio improvements and achieving associated cost savings, and (3) evaluate Office of Management and Budget's (OMB) plans to improve the PortfolioStat process.

To determine the status of agency efforts to implement key PortfolioStat actions, we obtained and analyzed policies, action plans, PortfolioStat briefing slides, status reports, agency communications to OMB, and other documentation relative to the key requirements of the Portfolio initiative outlined in OMB's 2012 memorandum[1] from each of the 26 federal agencies in our review.[2] These requirements included (1) designating a lead for the initiative; (2) completing a high-level IT portfolio survey; (3) establishing a commodity IT baseline; (4) holding a PortfolioStat session; (5) submitting a final plan to consolidate commodity IT; (6) migrating at least two duplicative commodity IT services by December 31, 2012; (7) and documenting lessons learned. For the final plan to consolidate commodity IT, we reviewed agency plans to determine whether each element required in the plan was fully addressed. A "partially" rating was given if the plan addressed a portion but not all of the information required in the element. In addition, we obtained a briefing book which OMB provided to the agencies that, among other things, summarized the agencies' commodity IT baseline data. We assessed the reliability of OMB's reporting of these data through interviews with OMB officials regarding their processes for compiling the briefing books and used the briefing books to describe the federal investment in commodity IT at the time of the 2012 PortfolioStat. We also assessed the reliability of agencies' commodity IT baseline data by reviewing the processes agencies described they had in place to ensure that all investments were captured in the baseline. We identified issues with the reliability of the

---

[1]OMB, *Memorandum for the Heads of Executive Departments and Agencies: Implementing PortfolioStat*, M-12-10 (Mar. 30, 2012).

[2]These agencies are the Departments of Agriculture, Commerce, Defense, Education, Energy, Health and Human Services, Homeland Security, Housing and Urban Development, the Interior, Justice, Labor, State, Transportation, the Treasury, and Veterans Affairs; the Environmental Protection Agency, General Services Administration, National Aeronautics and Space Administration, National Archives and Records Administration, National Science Foundation, Office of Personnel Management, Small Business Administration, Social Security Administration, U.S. Agency for International Development, U.S. Army Corps of Engineers, and the U.S. Nuclear Regulatory Commission.

agencies' commodity IT baseline data and have highlighted these issues throughout this report, as appropriate.

For objective two, we selected five agencies with (1) high fiscal year IT expenditure levels (based on information reported on the OMB's IT dashboard); (2) a mix of varying IT and CIO organizational structures (centralized vs. decentralized); and (3) a range of investment management maturity levels based on knowledge gathered from prior work and reported results of PortfolioStat sessions. In addition, to the extent possible, we avoided selecting projects that were the subject of another engagement underway. The agencies selected are the Departments of Agriculture, Defense, the Interior, the Treasury, and Veterans Affairs. To evaluate the selected agencies' plans for making portfolio improvements and achieving associated cost savings, we obtained and analyzed agencies' action plans to consolidate commodity IT, and other relevant documentation, and interviewed relevant agency officials to compile a list of planned portfolio improvements and determine the processes agencies used to identify these portfolio improvements. We determined the extent to which these processes included using (1) the agency enterprise architecture and (2) a valuation model, which OMB recommended in its guidance to assist in analyzing portfolio information and developing action plans. In addition, we assessed the reliability of the cost savings and avoidance estimates by obtaining and analyzing the support for the estimates for the two efforts that were to be migrated by December 2012 and the two efforts with the highest anticipated savings between fiscal years 2013 and 2015. Based on the results of our analysis, we found the data to be sufficiently reliable given the way they are reported herein.

To evaluate OMB's plans for making PortfolioStat improvements, we reviewed PortfolioStat guidance for fiscal year 2013[3] and interviewed OMB officials to compile a list of planned improvements. In addition, we analyzed the information obtained from our sources and the results of our analyses for our first two objectives to determine whether OMB's plans for improving PortfolioStat addressed the issues we identified.

---

[3]OMB, *Memorandum for the Heads of Executive Departments and Agencies: Fiscal Year 2013 PortfolioStat Guidance: Strengthening Federal IT Portfolio Management*, M-13-09 (Mar. 27, 2013).

We conducted this performance audit from October 2012 to November
2013 in accordance with generally accepted government auditing
standards. Those standards require that we plan and perform the audit to
obtain sufficient, appropriate evidence to provide a reasonable basis for
our findings and conclusions based on our audit objectives. We believe
that the evidence obtained provides a reasonable basis for our findings
and conclusions based on our audit objectives.

Appendix II: Agencies' Commodity IT Migration Efforts

The table below lists the commodity IT efforts for migration to shared services agencies identified in their action plan.

Agen__	__ffort__ for __igration to __are__ ervi__e__
Agriculture	Cellular Phone Contract Consolidation
	Enterprise Contracts for Standardized Security Products and Services Plan
Commerce	Cyber Security Assessment and Management Re-host at the Department of Justice
	National Oceanic and Atmospheric Administration Consolidated National Service Desk
	Endpoint Security Blanket Purchase Agreement
	Shared Services Freedom of Information Act Module Portal
Defense	The Unclassified Information Sharing Service/All Partner Access Network
	The General Fund Enterprise Business System
Education	Information Collection Request, Review and Approval System
	Kronos Time & Attendance
Energy	Electronic Capital Planning and Investment Control (eCPIC) Migration to General Services Administration Cloud Environment
	Energy Services Online Pilot Implementation
Environmental Protection Agency	Managed print services
	Strategic sourcing of end user computing
General Services Administration	Insite
	Contract writing module
Health and Human Services	Healthdata.gov Services
	Information Collection Request, Review and Approval Services
Homeland Security	WorkPlace as a Service
	Case and Relationship Management as a Service
Housing and Urban Development	Human Resource End to End Solution
	MicroStrategy Enterprise Licensing / Business Intelligence
Interior	Financial and Business Management System deployment 7
	Enterprise Forms System
Justice	Wireless
	Land Mobile Radio System Consolidation
Labor	Cloud e-mail
	Web-based collaboration tool

Agency	Effort for Migration to Shared Services
National Aeronautics and Space Administration	NASA Integrated Communications Services Consolidated Configuration Management System
	Consolidation of Adobe Lifecycle Reader Extension Service
National Archives and Records Administration	Moving website and census data to shared service provider
	Moving e-mail to shared services
National Science Foundation	Email to cloud
	Web Time & Attendance
Nuclear Regulatory Commission	Workforce Tracking Transformation system
	Entrance on Duty System
Office of Personnel Management	Help Desk consolidation
	IT asset inventory
Small Business Administration	Electronic Capital Planning and Investment Control (eCPIC) Portfolio Management tool (FESCOM Program Participant)
	FDOnline
Social Security Administration	Enterprise social media
	Geospatial architecture
State	Foreign Affairs Network
	Content publishing and delivery services
Transportation	Enterprise Messaging
	Systems and Managed Print Services and Multi-function printer devices
Treasury	Internet Payment Platform
	Do Not Pay
U.S. Agency for International Development	Google/Email
	Telecommunications and Computer Operations Center
U.S. Army Corps of Engineers	Electronic Capital Planning and Investment Control (eCPIC)
	USACE Learning Network
Veterans Affairs	Server Virtualization
	Eliminate Dedicated Fax Servers Consolidation

Source: GAO analysis of agency data.

The table below lists the commodity IT initiatives that agencies identified in the cost target templates provided to OMB in September 2012.

Initiative	Fi□al □ear □□1□ e□ti□ate□ □aving□ or □o□t avoi□an□e	Fi□al □ear □□14 e□ti□ate□ □aving□ or □o□t avoi□an□e	Fi□al □ear □□15 e□ti□ate□ □aving□ or □o□t avoi□an□e	□otal e□ti□ate□ □aving□ or □o□t avoi□an□e
				□ollar□ in □illion□ □ro□n□e□□
Agri□lt□re				
Infrastructure Consolidation	$61.10	$64.80	$78.30	$204.20
Tier 1 Help Desk	1.21	1.21	1.21	3.63
Geo Spatial Consolidation	10.53	20.60	30.03	61.16
□otal re□orte□ □aving□ an□ □o□t avoi□an□e	**7□□□4**	**□6□61**	**1□□54**	**□6□□□□**
□o□□er□e				
Desktop/Laptop Management	1.20	1.20	1.20	3.60
Several Data Center Consolidation Activities	0.00	5.40	29.30	34.70
Reduce total number of computers, use Commerce PC purchase contract to get discount.	0.38	0.38	0.38	1.14
National Oceanic and Atmospheric Administration National Service Desk	1.80	1.80	1.80	5.40
Enterprise-Wide IT Security Assessment and Authorization	1.80	1.80	1.80	5.40
Human Resources Management System	3.01	6.02	6.63	15.66
National Institute of Standards and Technology Cloud Initiatives	0.00	0.00	0.00	0.00
Voice over Internet Protocol	0.00	0.00	1.90	1.90
□otal re□orte□ □aving□ an□ □o□t avoi□an□e	**□□1□**	**16□6□**	**4□□□1**	**67□□□**
□efen□e[a]				
Branch Services Consolidation of Commodity IT Components and Applications	n.d.[b]	n.d.	n.d.	n.d.
Mobile Device Strategy	n.d.	n.d.	n.d.	n.d.
Next Generation End User Devices	n.d.	n.d.	n.d.	n.d.
Multi-level Security Domain Thin Client Solutions	n.d.	n.d.	n.d.	n.d.
Defense Enterprise Software Initiative	n.d.	n.d.	n.d.	n.d.
Consolidation Procurement of Commodity IT Hardware Purchases	n.d.	n.d.	n.d.	n.d.
Green IT	n.d.	n.d.	n.d.	n.d.
Unclassified Information Sharing Service / All Partner Access Network	n.d.	n.d.	n.d.	n.d.
Consolidate Security Infrastructure	n.d.	n.d.	n.d.	n.d.
Consolidate NetOps Centers	n.d.	n.d.	n.d.	n.d.
Implement Cross Domain Solution as Enterprise Service	n.d.	n.d.	n.d.	n.d.

Initiative	Fiscal Year 2013 estimated savings or cost avoidance	Fiscal Year 2014 estimated savings or cost avoidance	Fiscal Year 2015 estimated savings or cost avoidance	Total estimated savings or cost avoidance
Extended Joint Networks Over Satellite Communications	n.d.	n.d.	n.d.	n.d.
Video Over Internet Protocol Enterprise Service	n.d.	n.d.	n.d.	n.d.
Joint Enterprise Network	n.d.	n.d.	n.d.	n.d.
Defense Red Switch Network Rationalization	n.d.	n.d.	n.d.	n.d.
Data Center Consolidation	n.d.	n.d.	n.d.	n.d.
Computing Infrastructure and Services Optimization	n.d.	n.d.	n.d.	n.d.
Cloud Computing	n.d.	n.d.	n.d.	n.d.
Service Desk Consolidation	n.d.	n.d.	n.d.	n.d.
Enterprise Messaging and Collaboration Services	n.d.	n.d.	n.d.	n.d.
Identify and Access Management Services	n.d.	n.d.	n.d.	n.d.
Enterprises Services – Identify and Access Management	n.d.	n.d.	n.d.	n.d.
Streamline Records Management	n.d.	n.d.	n.d.	n.d.
Defense Interoperability with Mission Partners	n.d.	n.d.	n.d.	n.d.
General Fund Enterprise Business System	n.d.	n.d.	n.d.	n.d.
Common Business Process Foundation	n.d.	n.d.	n.d.	n.d.
Total reported savings and cost avoidance				5
Education				
Information Collection Request, Review and Approval System	0.15	0.15	0.15	0.45
Kronos Time & Attendance	0.56	0.33	0.33	1.22
ABLEDATA	0.08	0.00	0.00	0.08
cVent Event Solutions	0.00	0.01	0.01	0.02
Fund for the Improvement of PostSecondary Education Dissemination and Grants Database	0.03	0.03	0.03	0.08
Improving Program Performance	0.03	0.03	0.03	0.08
Literacy Information and Communication System Technical Services	0.11	0.11	0.11	0.32
Migrant Student Information Exchange	0.36	0.37	0.37	1.10
National Rehabilitation Information Center	0.02	0.03	0.03	0.07
Records Exchange Advice, Communication and Technical Support	0.00	0.00	0.00	0.00
Presidential Scholars Program	0.02	0.02	0.02	0.06
Asia Pacific Economic Cooperation Websites	0.00	0.00	0.00	0.01
Doing What Works Website	0.05	0.05	0.05	0.14
Education Pubs	0.03	0.03	0.03	0.10
eService Center	0.01	0.01	0.01	0.04

Initiative	Fiscal Year 2013 estimated savings or cost avoidance	Fiscal Year 2014 estimated savings or cost avoidance	Fiscal Year 2015 estimated savings or cost avoidance	Total estimated savings or cost avoidance
Office of the Chief Financial Officer Grants Information Award Database Internet Site	0.01	0.01	0.01	0.02
Enterprise Intranet (connectED)	0.11	0.11	0.11	0.32
Education Web	0.48	0.36	0.36	1.20
Total reported savings and cost avoidance	1.06	1.06	1.06	5.06
Energy				
Commodity IT Contract Consolidation	2.12	2.12	2.12	6.35
Utilization of Energy Master Contract	0.07	0.08	0.08	0.23
Enhanced Connectivity for Telework and Travel	0.03	0.03	0.03	0.08
Public Key Infrastructure Migration to Shared Service Provider	0.90	0.90	0.90	2.70
Email Consolidation	0.24	0.24	0.24	0.72
Collaboration Tools Consolidation (Microsoft SharePoint)	0.40	0.40	0.60	1.40
IT Security	0.00	0.36	0.36	0.72
Document Management Consolidation	0.15	0.18	0.22	0.55
Migration on-premise Exchange Services into Cloud 365 offering	0.00	0.30	0.30	0.60
Energy.gov Renewal Project	1.00	2.00	3.00	6.00
Rocky Mountain Oilfield Testing Center - Commodity IT Full Time Equivalent Reduction	0.23	0.23	0.46	0.92
Network Infrastructure	0.20	0.20	0.20	0.60
Server and Storage Consolidation	0.40	0.40	0.40	1.20
eCPIC Migration to General Services Administration Cloud Environment	0.10	0.13	0.13	0.35
Implement CISCO Unified Communication & Collaboration	1.90	2.30	2.60	6.80
ITSM Replacement of Office of the Chief Information Officer Remedy systems with ServiceNow	0.48	0.50	0.52	1.50
Total reported savings and cost avoidance	8.11	10.36	12.15	30.7
Environmental Protection Agency				
Email (Software as a Service)	0.00	0.00	0.00	0.00
Collaboration Tools (Software as a Service)	0.00	0.00	0.00	0.00
Total reported savings	0.00	0.00	0.00	0.00
General Services Administration				
External Services Branch Consolidation	1.88	1.96	5.88	9.72
Identity Credentials and Access Management	.40	.97	1.55	2.92
Insite	(.16)	.36	.36	.56
Total reported savings and cost avoidance	2.11	3.06	7.79	13.00

Initiative	Fiscal Year 2013 estimated savings or cost avoidance	Fiscal Year 2014 estimated savings or cost avoidance	Fiscal Year 2015 estimated savings or cost avoidance	Total estimated savings or cost avoidance
Health and Human Services				
Human Resources Consolidation and Migration to Shared Service Provider	0.00	TBD	TBD	0.00
Email Migration to E-mail as a Service Provider	0.00	TBD	TBD	0.00
Total reported savings and cost avoidance	0.00	0.00	0.00	0.00
Homeland Security				
Email as a Service	18.83	48.40	48.40	115.63
Workplace as a Service	73.39	153.80	153.80	380.99
Networx Annual Cost Avoidance	85.60	85.60	85.60	256.80
Enterprise Licensing Agreements	125.33	125.33	125.33	375.99
Wireless	1.00	1.00	1.00	3.00
Homeland Security Data Center Consolidation	24.40	24.40	24.40	73.20
Identify, Credential, and Access Management	2.40	2.40	2.40	7.20
Transportation Security Administration Transportation Threat Assessment and Credentialing/Infrastructure Modernization Program	12.00	12.00	12.00	36.00
Office of the Chief Information Officer/Enterprise System Development Office – Expected reduction in HQ operational expenditures	0.00	4.00	4.00	8.00
Web Content Management as a Service	0.00	18.00	18.00	36.00
Enterprise Content Delivery as a Service Cost Avoidance	6.00	6.00	6.00	18.00
SharePoint as a Service	2.16	2.16	2.16	6.48
GeoSpatial (Enterprise IT Services)	16.75	16.75	16.75	50.25
Homeland Security Information Network	1.41	1.41	1.41	4.22
Case and Relationship Management as a Service	0.00	0.00	0.00	0.00
Total reported savings and cost avoidance	369.27	501.65	501.65	1,371.77
Housing and Urban Development				
Human Resource End to End Solution	0.24	0.24	0.23	0.71
Standard Business Intelligence	0.31	0.31	0.31	0.93
Email to the Cloud	1.32	1.32	1.32	3.96
Microsoft Enterprise Software Licensing	1.00	1.00	1.00	3.00
Total reported savings and cost avoidance	2.87	2.87	2.86	8.60
Interior				
Networx	7.30	7.30	7.30	21.90
Circuit Consolidation	0.90	0.90	0.90	2.70
Cloud Email and Collaboration Services	4.52	4.52	4.52	13.56

Initiative	Fiscal Year 2013 estimated savings or cost avoidance	Fiscal Year 2014 estimated savings or cost avoidance	Fiscal Year 2015 estimated savings or cost avoidance	Total estimated savings or cost avoidance
Enterprise eArchive System part of eMail Enterprise Records and Document Management System	3.80	3.80	3.80	11.40
Financial and Business Management System deployment 7&8	1.30	4.13	0.00	5.43
Enterprise Forms System	2.30	2.30	2.30	6.90
Total reported savings and cost avoidance	**21**	**45**	**1**	**61**
Justice				
Consolidation of Classified Processing Services	n.d.	n.d.	n.d.	n.d.
Consolidating number of email systems	n.d.	n.d.	n.d.	n.d.
Human Resource System initiatives	n.d.	n.d.	n.d.	n.d.
Web Time and Attendance Cloud Solution	n.d.	n.d.	n.d.	n.d.
Wireless contract consolidation	n.d.	n.d.	n.d.	n.d.
Justice Management Division Mobility-Virtual Private Network	n.d.	n.d.	n.d.	n.d.
Consolidation of Justice Land Mobile Radio Systems	n.d.	n.d.	n.d.	n.d.
Monitoring at two security operations centers	n.d.	n.d.	n.d.	n.d.
Telecommunications savings initiatives	n.d.	n.d.	n.d.	n.d.
Bureau of Alcohol, Tobacco, Firearms and Explosives Unified Communications	n.d.	n.d.	n.d.	n.d.
Strategic sourcing (contract escalations)	n.d.	n.d.	n.d.	n.d.
Network Delivery Order for CISCO services	n.d.	n.d.	n.d.	n.d.
Total reported savings and cost avoidance				**5**
Labor				
DOLNet Network Infrastructure consolidation	0.00	0.00	0.00	0.00
Managed Trusted Internet Protocol Service	0.00	0.00	0.00	0.00
Federal Data Center Consolidation Initiative	0.00	0.34	0.67	1.01
E Grants-Related Federal Financial Assistance	0.00	0.00	0.00	0.00
Cloud Email	0.00	1.40– 2.70	1.40– 2.70	2.80–5.40
Labor Human Resources Works	0.00	0.00	0.00	0.00
HR Works Formerly Human Resources Line of Business Shared Service Center	0.00	0.00	0.00	0.00
Homeland Security Presidential Directive 12	0.50	1.00	3.00	4.50
Contractor Personnel System	0.50	0.75	1.00	2.25
New Core Financial Management System	0.00	0.00	0.00	0.00
Web content management and hosting consolidation	0.00	2.00	2.79	4.79
Web content HTML editors	0.75	3.10	3.10	6.95
Total reported savings and cost avoidance	**1.75**	**8.59–9.89**	**11.96–13.26**	**22.30–24.90**

Initiative	Fiscal Year 2013 estimated savings or cost avoidance	Fiscal Year 2014 estimated savings or cost avoidance	Fiscal Year 2015 estimated savings or cost avoidance	Total estimated savings or cost avoidance
National Aeronautics and Space Administration				
Consolidated Corporated Operations Network Center	0.00	1.66	2.21	3.87
IT Security	0.00	0.00	0.00	0.00
Collaboration Tools	0.00	0.00	0.00	0.00
Agency Consolidated End User Services	18.85	19.23	19.61	57.69
NASA Integrated Communication Services Consolidated Configuration Management System	0.00	2.24	2.99	5.23
Consolidation of Adobe Life Cycle Reader Extension Service	0.17	0.16	0.16	0.49
Total reported savings and cost avoidance	19.02	23.29	24.97	67.28
National Archives and Records Administration				
Human Resources Service Center Migration	-0.02	-0.02	-0.02	-0.06
Web Hosting of the 1940 Decennial Census	0.00	0.00	3.29	3.29
Web Hosting of Archives.gov	0.00	0.00	3.20	3.20
Total reported savings and cost avoidance	-0.02	-0.02	6.47	6.43
National Science Foundation				
Email to Cloud	0.27	0.24	0.24	0.75
Web Time & Attendance	0.00	0.00	0.00	0.00
Total reported savings and cost avoidance	0.27	0.24	0.24	0.75
Nuclear Regulatory Commission				
IPiSS PIV Legacy Modernization Savings	0.00	0.92	0.00	0.92
Security and Forensics Efficiencies	0.00	0.34	0.00	0.34
Streamline Multiple Web Content Publications through centralized Web content solution	0.00	1.25	0.00	1.25
Electronic Information Exchange	0.00	0.02	0.00	0.02
Data Center Outsourcing	0.00	4.06	0.00	4.06
Consolidating Video Conferencing	0.00	0.50	0.00	0.50
Famis Rehosting / E-Travel new contract	0.00	2.42	0.00	2.42
HR Management System	0.00	0.08	0.00	0.08
Total reported savings and cost avoidance	0.00	9.60	0.00	9.60
Office of Personnel Management				
Mobile Device and Service Plan Consolidation	0.00	1.50	0.00	1.50
Printer Consolidation	0.00	2.00	0.00	2.00
Total reported savings and cost avoidance	0.00	3.50	0.00	3.50

Initiative	Fiscal Year 2013 estimated savings or cost avoidance	Fiscal Year 2014 estimated savings or cost avoidance	Fiscal Year 2015 estimated savings or cost avoidance	Total estimated savings or cost avoidance
Small Business Administration				
Data Center migration to shared center	0.00	0.00	0.00	0.00
SBA Mobility	0.16	0.16	0.00	0.31
Learning Management System aka Integrated Talent Management System = LMS (Learning Management System) + PM (Performance Management) $265k	0.06	0.03	0.00	0.09
WFA aka Workforce Analytics or Workforce Planning $535k	0.06	0.03	0.00	0.09
eCPIC Portfolio Management Tool (FESCOM Program Participant)	0.15	0.15	0.00	0.30
FDonline	0.00	0.00	0.00	0.00
Total reported savings and cost avoidance	**.43**	**.37**	**.00**	**.79**
Social Security Administration				
Mainframe hardware and maintenance, TN3270 rollout to the field	0.00	0.68	0.00	0.68
SSA workforce consolidation and improvement	6.65	6.65	6.65	19.95
Disability Case Processing System implementation	0.00	0.31	1.55	1.86
Enterprise desktop refresh	0.00	58.74	0.00	58.74
Consolidation of Open Systems	0.58	0.58	0.58	1.74
Mainframe hardware and maintenance	0.00	22.35	0.00	22.35
Software maintenance	0.00	60.84	0.00	60.84
Total reported savings and cost avoidance	**7.23**	**150.15**	**8.78**	**166.16**
State				
Strategic Sourcing Initiative	0.00	5.00	1.00	6.00
Global IT Modernization	0.00	5.00	1.00	6.00
Enterprise licensing software	0.00	5.00	1.00	6.00
Printer Managed Service	0.00	5.00	3.00	8.00
Total reported savings and cost avoidance	**.00**	**.00**	**6.00**	**26.00**
Transportation				
Federal Motor Carrier Safety Administration SharePoint Server Migration	0.00	0.00	TBD	0.00
Pipeline and Hazardous Safety Materials Administration SharePoint Migration	0.00	0.04	TBD	0.04
Transportation Common Operating Environment SharePoint 2007 Decommissioning	0.00	0.07	TBD	0.07
Federal Railroad Administration Correspondence Control Management System Migration to Departmental Solution	0.00	0.08	TBD	0.08
Transportation Common Operating Environment SharePoint 2010 Migration to EMS Solution	0.00	0.50	TBD	0.50

Initiative	Fiscal Year 2013 estimated saving or cost avoidance	Fiscal Year 2014 estimated saving or cost avoidance	Fiscal Year 2015 estimated saving or cost avoidance	Total estimated saving or cost avoidance
Pipeline and Hazardous Safety Materials Administration Migration to Shared Solution	0.00	0.09	TBD	0.09
National Highway Transportation Safety Administration Transition to the DOT Common Operating Environment	0.00	0.19	TBD	0.19
Federal Motor Carrier Safety Administration Migration to Enterprise Web	0.00	0.38	TBD	0.38
Transportation Common Operating Environment Legacy Migration to Cloud	0.00	0.50	TBD	0.50
Research and Innovative Technology Administration Web Deployment in the Cloud	0.00	0.10	TBD	0.10
Transportation IBM BigFix	0.00	TBD	TBD	0.00
Federal Highway Administration National Highway Institute Web Portal & Course Management Consolidation to Reduce IT Security	0.00	0.11	TBD	0.11
Federal Highway Administration IT Security Process Improvements	0.00	2.54	TBD	2.54
Transportation COE Wireless Reinvest in DOT-wide Solution	0.00	1.06	TBD	1.06
Federal Railroad Administration Mobile Workforce Initiative Wireless Services	0.00	0.17	0.00	0.17
Transportation Common Operating Environment Migration from Current Host Provider	0.00	2.24	TBD	2.24
Federal Motor Carrier Safety Administration Virtualization of Server Environment	0.00	0.16	TBD	0.16
National Highway Transportation Safety Administration Teleprocessing and Timesharing Services for the National Driver Register Program Cloud Hosting	0.00	2.46	TBD	2.46
National Highway Transportation Safety Administration Common IT Services Transition to Cloud-Based Hosting	0.00	3.13	TBD	3.13
Transportation Common Operating Environment De-Duplication Project	0.84	0.85	TBD	1.69
Research and Innovative Technology Administration Server Migration and Consolidation Initiative	0.17	0.40	TBD	0.57
Federal Aviation Administration Email	13.40	12.00	TBD	25.40
Transportation Common Operating Environment Email	0.00	4.50	TBD	4.50
Transportation Headquarters Migration to FAA SAVES MFP Contract	0.00	5.00	TBD	5.00
Total reported savings and cost avoidance	**14.41**	**36.57**	**0.00**	**50.96**
Treasury				
Internet Payment Platform	5.00	25.00	35.00	65.00
DoNotPay	19.03	42.75	66.00	127.78
Total reported savings and cost avoidance	**24.03**	**67.75**	**101.00**	**192.78**

Initiative	Fiscal Year 2013 estimated savings or cost avoidance	Fiscal Year 2014 estimated savings or cost avoidance	Fiscal Year 2015 estimated savings or cost avoidance	Total estimated savings or cost avoidance
U.S. Agency for International Development				
Google/Email	1.17	1.47	2.29	4.93
Single Device	0.10	0.10	0.10	0.31
Telecommunications and Computer Operations Center	0.63	2.13	4.83	7.59
Public-facing Website Consolidation	1.54	1.54	1.54	4.62
Total reported savings and cost avoidance	3.44	5.24	8.76	17.44
U.S. Army Corps of Engineers				
eCPIC	0.00	0.00	0.00	0.00
USACE Learning Network	0.50	0.50	0.50	1.50
Total reported savings and cost avoidance	0.50	0.50	0.50	1.50
Veterans Affairs				
Server Virtualization	0.42	0.75	1.10	2.28
Eliminate Dedicated Fax Servers Consolidation	0.36	9.72	45.00	55.08
Microsoft Enterprise Licensing Agreement	40.00	40.00	40.00	120.00
Standardize Spend Planning and Consolidation Contracts	2.23	0.24	0.12	2.59
Vants Via Voice Over Internet Protocol	0.60	1.80	3.00	5.40
Vista Data Feeds	5.00	0.00	0.00	5.00
One CPU Policy	5.57	0.00	0.00	5.57
Total reported savings and cost avoidance	54.18	52.51	89.22	195.91
Total reported savings and cost avoidance (all agencies)[d]	61,097.7	1,000,145.14 1,000,485.14	654,365.6 656,046.6	5,774,686.6 7,774,746.6

Source: GAO analysis of agency data.

[a]Defense did not provide information on the number of opportunities or potential cost savings to OMB in the cost target template. The information was obtained from the department's final action plan and other referenced documentation. Defense reported savings for fiscal years 2012-2015 and did not break down cost savings by initiative.

[b]n.d.—no data.

[c]Justice did not provide information on the number of opportunities or potential cost savings to OMB in the cost target template. The information was obtained from the department's final action plan. Justice reported savings for fiscal year 2015 and did not break down cost savings by initiative.

[d]Numbers may not add up due to rounding.

Appendix IV: Recommendations to Departments and Agencies

Agriculture	To improve the department's implementation of PortfolioStat, we recommend that the Secretary of Agriculture direct the CIO to take the following four actions:

- Develop a complete commodity IT baseline.

- In future reporting to OMB, fully describe the following PortfolioStat Action plan elements: (1) *consolidate commodity IT spending under the agency CIO* and (2) *establish criteria for wasteful, low-value, or duplicative investments.*

- As the department finalizes and matures its valuation methodology, utilize this process to identify whether there are additional opportunities to reduce duplicative, low-value, or wasteful investments.

- Develop support for the estimated savings for fiscal years 2013 through 2015 for the Cellular Phone Contract Consolidation, IT Infrastructure Consolidation/Enterprise Data Center Consolidation, and Geospatial Consolidation initiatives.

Commerce	To improve the department's implementation of PortfolioStat, we recommend that the Secretary of Commerce direct the CIO to take the following two actions:

- Reflect 100 percent of information technology investments in the department's enterprise architecture.

- Develop a complete commodity IT baseline.

Defense	To improve the department's implementation of PortfolioStat, we recommend that the Secretary of Defense direct the CIO to take the following three actions:

- Develop a complete commodity IT baseline.

- In future reporting to OMB, fully describe the following PortfolioStat action plan element: *consolidate commodity IT spending under the agency CIO.*

- Obtain support from the relevant component agencies for the estimated savings for fiscal years 2013 to 2015 for the data center consolidation, enterprise software purchasing, and General Fund Enterprise Business System initiatives.

In addition, to improve the U.S. Army Corps of Engineers' implementation of PortfolioStat, we recommend that the Secretary of Defense direct the Secretary of the Army to take the following two actions:

- In future reporting to OMB, fully describe the following PortfolioStat action plan elements: (1) *consolidate commodity IT spending under the agency CIO*; (2) *target duplicative systems or contracts that support common business functions for consolidation*; (3) *establish criteria for identifying wasteful, low-value, or duplicative investments*; and (4) *establish a process to identify these potential investments and a schedule for eliminating them from the portfolio.*.

- Report on the agency's progress in consolidating eCPIC to a shared service as part of the OMB integrated data collection quarterly reporting until completed.

Energy

To improve the department's implementation of PortfolioStat, we recommend that the Secretary of Energy direct the CIO to take the following action:

- In future reporting to OMB, fully describe the following PortfolioStat action plan elements: (1) *consolidate commodity IT spending under the agency CIO* and (2) *establish criteria for identifying wasteful, low-value, or duplicative investments*.

Environmental Protection Agency

To improve the agency's implementation of PortfolioStat, we recommend that the Administrator of the Environmental Protection Agency direct the CIO to take the following three actions:

- Develop a complete commodity IT baseline.

- In future reporting to OMB, fully describe the following PortfolioStat action plan elements: (1) *consolidate commodity IT spending under the agency CIO*; (2) *establish targets for commodity IT spending reductions and deadlines for meeting those targets*; and (3) *establish criteria for identifying wasteful, low-value, or duplicative investments*.

- Report on the agency's progress in consolidating the managed print services and strategic sourcing of end user computing to shared services as part of the OMB integrated data collection quarterly reporting until completed.

General Services Administration

To improve the agency's implementation of PortfolioStat, we recommend that the Administrator of the General Services Administration direct the CIO to take the following action:

- Report on the agency's progress in consolidating the contract writing module to a shared service as part of the OMB integrated data collection quarterly reporting until completed.

Health and Human Services

To improve the department's implementation of PortfolioStat, we recommend that the Secretary of Health and Human Services direct the CIO to take the following action:

- In future OMB reporting, fully describe the following PortfolioStat action plan element: *consolidate commodity IT spending under the agency CIO.*

Housing and Urban Development

To improve the department's implementation of PortfolioStat, we recommend that the Secretary of Housing and Urban Development direct the CIO to take the following three actions:

- Develop a complete commodity IT baseline.

- In future reporting to OMB, fully describe the following PortfolioStat action plan element: *establish criteria for identifying wasteful, low-value, or duplicative investments.*

- Report on the department's progress in consolidating the HR End-to-End Performance Management Module to a shared service as part of the OMB integrated data collection quarterly reporting until completed.

Interior

To improve the department's implementation of PortfolioStat, we recommend that the Secretary of the Interior direct the CIO to take the following three actions:

- Develop a complete commodity IT baseline.

- In future reporting to OMB, fully describe the following PortfolioStat action plan element: *establish criteria for identifying wasteful, low-value, or duplicative investments.*

- Report on the department's progress in consolidating the Electronic Forms System component of the eMail Enterprise Records & Document Management System deployment 8 to a shared service as part of the OMB integrated data collection quarterly reporting until completed.

Justice

To improve the department's implementation of PortfolioStat, we recommend that the Attorney General direct the CIO to take the following two actions:

- Reflect 100 percent of information technology investments in the department's enterprise architecture.

- In future reporting to OMB, fully describe the following PortfolioStat action plan element: *establish targets for commodity IT spending reductions and deadlines for meeting those targets.*

Labor

To improve the department's implementation of PortfolioStat, we recommend that the Secretary of Labor direct the CIO to take the following three actions:

- Develop a complete commodity IT baseline.

- In future reporting to OMB, fully describe the following PortfolioStat action plan elements: (1) *consolidate commodity IT spending under the agency CIO* and (2) *establish targets for commodity IT spending reductions and deadlines for meeting those targets.*

- Report on the department's progress in consolidating the cloud e-mail services to a shared service as part of the OMB integrated data collection quarterly reporting until completed.

National Aeronautics and Space Administration

To improve the agency's implementation of PortfolioStat, we recommend that the Administrator of the National Aeronautics and Space Administration direct the CIO to take the following three actions:

- Reflect 100 percent of information technology investments in the agency's enterprise architecture.

- In future reporting to OMB, fully describe the following PortfolioStat action plan elements: (1) *consolidate commodity IT spending under the agency CIO*; (2) *target duplicative systems or contracts that support common business functions for consolidation*; (3) *establish criteria for identifying wasteful, low-value, or duplicative investments*; and (4) *establish a process to identify these potential investments and a schedule for eliminating them from the portfolio.*

- Report on the agency's progress in consolidating the NASA Integrated Communications Services Consolidated Configuration Management System to a shared service as part of the OMB integrated data collection quarterly reporting until completed.

National Archives and Records Administration

To improve the agency's implementation of PortfolioStat, we recommend that the Archivist of the United States direct the CIO to take the following action:

- In future reporting to OMB, fully describe the following PortfolioStat action plan elements: (1) *consolidate commodity IT spending under the agency CIO*; (2) *target duplicative systems or contracts that support common business functions for consolidation*; (3) *establish criteria for identifying wasteful, low-value, or duplicative investments*; and (4) *establish a process to identify these potential investments and a schedule for eliminating them from the portfolio.*

National Science Foundation

To improve the agency's implementation of PortfolioStat, we recommend that the Director of the National Science Foundation direct the CIO to take the following action:

- In future reporting to OMB, fully describe the following PortfolioStat action plan elements: (1) *consolidate commodity IT spending under the agency CIO* and (2) *establish criteria for identifying wasteful, low-value, or duplicative investments.*

Nuclear Regulatory Commission

To improve the agency's implementation of PortfolioStat, we recommend that the Chairman of the U.S. Nuclear Regulatory Commission direct the CIO to take the following two actions:

- Develop a complete commodity IT baseline.

	In future reporting to OMB, fully describe the following PortfolioStat action plan elements: (1) *consolidate commodity IT spending under the agency CIO*; (2) *establish targets for commodity IT spending reductions and deadlines for meeting those targets*; (3) *target duplicative systems or contracts that support common business functions for consolidation;* and (4) *establish a process to identify these potential investments and a schedule for eliminating them from the portfolio.*

Office of Personnel Management

To improve the agency's implementation of PortfolioStat, we recommend that the Director of the Office of Personnel Management direct the CIO to take the following three actions:

- Develop a complete commodity IT baseline.

- In future reporting to OMB, fully describe the following PortfolioStat action plan elements: (1) *move at least two commodity IT areas to shared services* and (2) *target duplicative systems or contracts that support common business functions for consolidation.*

- Report on the agency's progress in consolidating the help desk consolidation and IT asset inventory to shared services as part of the OMB integrated data collection quarterly reporting until completed.

Small Business Administration

To improve the agency's implementation of PortfolioStat, we recommend that the Administrator of the Small Business Administration direct the CIO to take the following two actions:

- Develop a complete commodity IT baseline.

- In future reporting to OMB, fully describe the following PortfolioStat action plan elements: (1) *consolidate commodity IT spending under the agency CIO*; (2) *establish targets for commodity IT spending reductions and deadlines for meeting those targets*; (3) *target duplicative systems or contracts that support common business functions for consolidation;* and (4) *establish a process to identify those potential investments and a schedule for eliminating them from the portfolio.*

Social Security Administration

To improve the agency's implementation of PortfolioStat, we recommend that the Commissioner of the Social Security Administration direct the CIO to take the following two actions:

- Develop a complete commodity IT baseline.

- Report on the agency's progress in consolidating the geospatial architecture to a shared service as part of the OMB integrated data collection quarterly reporting until completed.

State

To improve the department's implementation of PortfolioStat, we recommend that the Secretary of State direct the CIO to take the following three actions:

- Reflect 100 percent of information technology investments in the department's enterprise architecture.

- In future reporting to OMB, fully describe the following PortfolioStat action plan elements: (1) *consolidate commodity IT spending under the agency CIO*; (2) *establish targets for commodity IT spending reductions and deadlines for meeting those targets*; (3) *move at least two commodity IT areas to shared services*; (4) *target duplicative systems or contracts that support common business functions for consolidation*; and (5) *establish a process to identify those potential investments and a schedule for eliminating them from the portfolio.*

- Report on the department's progress in consolidating the Foreign Affairs Network and content publishing and delivery services to shared services as part of the OMB integrated data collection quarterly reporting until completed.

Transportation

To improve the department's implementation of PortfolioStat, we recommend that the Secretary of Transportation direct the CIO to take the following two actions:

- In future reporting to OMB, fully describe the following PortfolioStat action plan elements: (1) *consolidate commodity IT spending under the agency CIO*; (2) *establish targets for commodity IT spending reductions and deadlines for meeting those targets*; (3) *target duplicative systems or contracts that support common business functions for consolidation*; and (4) *establish a process to identify those potential investments and a schedule for eliminating them from the portfolio.*

- Report on the department's progress in consolidating the Enterprise Messaging to shared services as part of the OMB integrated data collection quarterly reporting until completed.

Treasury	To improve the department's implementation of PortfolioStat, we recommend that the Secretary of the Treasury direct the CIO to take the following three actions:

- In future reporting to OMB, fully describe the following PortfolioStat action plan elements: (1) *consolidate commodity IT spending under the agency CIO* and (2) *establish criteria for identifying wasteful, low-value, or duplicative investments*.

- As the department finalizes and matures its enterprise architecture and valuation methodology, utilize these processes to identify whether there are additional opportunities to reduce duplicative, low-value, or wasteful investments.

- Develop support for the estimated savings for fiscal years 2013 to 2015 for the DoNotPay Business Center, Fiscal IT Data Center Consolidation and Business Process Management Status initiatives.

U.S. Agency for International Development	To improve the agency's implementation of PortfolioStat, we recommend that the Administrator of the U.S. Agency for International Development direct the CIO to take the following four actions:

- Reflect 100 percent of information technology investments in the agency's enterprise architecture.

- Develop a complete commodity IT baseline.

- In future reporting to OMB, fully describe the following PortfolioStat action plan elements: (1) *target duplicative systems or contracts that support common business functions for consolidation* and (2) *establish a process to identify those potential investments and a schedule for eliminating them from the portfolio*.

- Report on the agency's progress in consolidating the e-mail and Telecommunication and Operations Center to shared services as part of the OMB integrated data collection quarterly reporting until completed.

Veterans Affairs	To improve the department's implementation of PortfolioStat, we recommend that the Secretary of Veterans Affairs direct the CIO to take the following four actions:

- In future reporting to OMB, fully describe the following PortfolioStat action plan element: *target duplicative systems or contracts that support common business functions for consolidation.*

- Report on the department's progress in consolidating the dedicated fax servers to a shared service as part of the OMB integrated data collection quarterly reporting until completed.

- As the department matures its enterprise architecture process, make use of it, as well as the valuation model, to identify whether there are additional opportunities to reduce duplicative, low-value, or wasteful investments.

- Develop detailed support for the estimated savings for fiscal years 2013 to 2015 for the Server Virtualization, Eliminate Dedicated Fax Servers Consolidation, Renegotiate Microsoft Enterprise License Agreement, and one CPU policy initiatives.

Appendix V: Comments from the U.S. Department of Agriculture

USDA United States Department of Agriculture

Departmental
Management

Office of the Chief
Information Officer

1400 Independence
Avenue S.W.
Washington, DC
20250

David Powner
Director
Information Technology Management Issues
U.S. Government Accountability Office
441 G Street, N. W.
Washington, DC 20548

SEP 1 9 2013

Dear Mr. Powner:

The U.S. Department of Agriculture has reviewed the draft report GAO- Draft Report GAO-14-65, October 2013.

Thank you for the opportunity to respond to the GAO draft report. We concur with the content of the report and have no comments.

For additional information, please contact Christopher Wren, Office of the Chief Information Officer's audit liaison, at 202-260-0771.

Sincerely,

Cheryl L. Cook
Chief Information Officer

AN EQUAL OPPORTUNITY EMPLOYER

Appendix VI: Comments from the Department of Commerce

THE DEPUTY SECRETARY OF COMMERCE
Washington, D.C. 20230

September 25, 2013

Mr. David A. Powner
Director, Information Technology Management
United States Government Accountability Office
Washington, DC 20548

Dear Mr. Powner:

Thank you for the opportunity to comment on the draft report from the U.S. Government Accountability Office (GAO) titled "Information Technology: Additional OMB and Agency Actions Are Needed to Achieve Portfolio Savings" (GAO-14-65).

The Department of Commerce embraces the PortfolioStat initiative and we believe we have made significant strides to implement all the key provisions of both PortfolioStat 2012 and PortfolioStat 2013. Below are our comments regarding those areas within the draft report that relate specifically to the Commerce Department.

On page 11, the draft report states, "The Department of Commerce reported that its CIO only has explicit authority over major IT investments." We feel this to be an incorrect characterization of the authorities held by the Commerce CIO. On June 21, 2012, the Acting Secretary of Commerce issued a memorandum promulgating the Department's IT Portfolio Management Policy. That Policy states the following:

> *While funding generally resides with the Operating Units, pursuant to the CIO's oversight responsibilities, the CIO shall have discretion to review and approve IT investments and acquisitions during budget formulation and program execution.*

> *The CIO shall establish IT investment and/or acquisition review authorities (based on criteria that may include initial cost, annual cost, total lifecycle cost). Any DOC IT investment and/or acquisition that exceeds these thresholds shall be subject to review by the CIO via the Commerce IT Review Board (CITRB) or other mechanism identified by the CIO and/or the Chief Acquisition Officer.*

We feel there is nothing within the Acting Secretary's memorandum that restricts the CIO's authority to review and oversee major IT investments, and the authority to review any IT investment, whether major or nonmajor, is explicit in the memorandum.

Mr. David A. Powner
Page 2

We concur with the observation on page 13 of the draft report that 90 percent of Commerce's IT investments are reflected in the enterprise architecture.

We concur with the comment on page 16 of the draft report that Commerce could not ensure the completeness of our commodity IT baseline at the time the GAO conducted its study. Since then, however, we have established a significantly improved commodity IT baseline that is considerably more comprehensive, per the requirements established through the 2013 PortfolioStat process. The 2013 PortfolioStat process used for the compilation of this revised baseline (1) was developed by OMB, (2) includes a broader set of IT commodities than did our previous one, and (3) has been uniformly applied across the 26 agencies required to comply with OMB's memo for implementing the PortfolioStat initiative.

We concur with the portion of the "Recommendations to Departments and Agencies" on page 56 of the draft report, recommending that Commerce (1) reflect 100 percent of information technology investments in the department's enterprise architecture and (2) develop a complete commodity IT baseline.

Please contact Jerry Harper, Director, Office of IT Policy and Planning, at 202-482-0222 if you have questions regarding this response.

Sincerely,

Patrick Gallagher
Acting Deputy Secretary of Commerce

Appendix VII: Comments from the Department of Defense

DEPARTMENT OF DEFENSE
6000 DEFENSE PENTAGON
WASHINGTON, D.C. 20301-6000

CHIEF INFORMATION OFFICER

SEP 2 7 2013

Mr. David A. Powner,
Director, Information Technology Management Issues
U.S. Government Accountability Office
441 G Street, NW
Washington, DC 20548

Dear Mr. Powner,

This is the Department of Defense response to the GAO Draft Report, GAO-14-65, "INFORMATION TECHNOLOGY: Additional OMB and Agency Actions are needed to Achieve Portfolio Savings," dated August 29, 2013 (GAO Code 311280).

Our comments to the draft report are attached. My point of contact is Mr. Kevin Garrison, 571-372-4473, kevin.garrison1.civ@mail.mil.

Sincerely,

David L. DeVries
Deputy Chief Information Officer for
Information Enterprise

Attachment:
As stated

GAO DRAFT REPORT DATED AUGUST 29, 2013
GAO-14-65 (GAO CODE 311280)

**"INFORMATION TECHNOLOGY: Additional OMB and Agency Actions are Needed to
Achieve Portfolio Savings"**

DEPARTMENT OF DEFENSE COMMENTS
TO THE GAO RECOMMENDATIONS

RECOMMENDATION 1: To improve the department's implementation of PortfolioStat, we
recommend that the Secretary of Defense direct the CIO to take the following action: Develop a
complete commodity IT baseline.

DoD RESPONSE: DoD partially concurs with this recommendation. DoD has efforts
underway, including the Joint Information Environment, to further refine the Department's
commodity IT baseline.

RECOMMENDATION 2: To improve the department's implementation of PortfolioStat, we
recommend that the Secretary of Defense direct the CIO to take the following action: Fully
describe the following PortfolioStat action plan element, consolidate commodity IT spending
under the agency CIO, in future OMB reporting.

DoD RESPONSE: DoD does not concur with this recommendation. The commodity IT
construct OMB has implemented in PortfolioStat does not work within current federated
management processes of the DoD. The Department operates as a decentralized organization,
and its federated decentralized approach is the same approach used for all major decision-making
processes within the Department. The DoD CIO has integrated and balanced the authorities
outlined in M-11-29 and legislative requirements contained in Title 10, Title 40 and Title 44 to
achieve improved IT operating efficiencies and organizational decision-making agility.

The federated management approach takes advantage of long standing and highly
effective financial and acquisition management processes, by using Department staff resources
efficiently thereby reducing duplication of efforts. Our approach also allows for leveraging the
extensive development, monitoring and control processes for both budget and acquisition that are
found in every Military Service, Agency and Component operations. DoD does agree that a
strategy, consistent with the intent of achieving better buying power and control of commodity
IT items, should be developed and implemented within the Department using existing authorities
and is in the process of implementing such a strategy.

OMB M-1 1-29, Chief Information Officer Authorities, http://www.whitehouse.gov/sites/default/files/omb/memoranda/2011/mll-29.pdf. Commodity IT
was defined as including services such as, "IT infrastructure (data centers, networks, desktop computers and mobile devices); enterprise IT systems
(email, collaboration tools, identity and access management, security, and web infrastructure); and business systems (finance, human resources, and
other administrative functions)."

RECOMMENDATION 3: To improve the department's implementation of PortfolioStat, we
recommend that the Secretary of Defense direct the CIO to take the following action: Obtain

GAO DRAFT REPORT DATED AUGUST 29, 2013
GAO-14-65 (GAO CODE 311280)

"INFORMATION TECHNOLOGY: Additional OMB and Agency Actions are Needed to
Achieve Portfolio Savings"

DEPARTMENT OF DEFENSE COMMENTS
TO THE GAO RECOMMENDATIONS

support from the relevant component agencies for the estimated savings for fiscal years 2013 to
2015 for the data center consolidation, enterprise software purchasing, and General Fund
Enterprise Business System initiatives.

DoD RESPONSE: DoD concurs with this recommendation. DoD already reports data center
consolidation savings to both OMB and Congress and will continue to realize savings from the
Enterprise Software Initiative, other strategic sourcing efforts, and the continuing
implementation of GFEBS.

RECOMMENDATION 4: To improve the U.S. Army Corps of Engineers' implementation of
PortfolioStat, we recommend that the Secretary of Defense direct the Secretary of the Army to
take the following action: Fully describe the following PortfolioStat action plan elements: (1)
consolidate commodity IT spending under the agency CIO; (2) target duplicative systems or
contracts that support common business functions for consolidation; (3) establish criteria for
identifying wasteful, low-value, or duplicative investments; and (4) establish a process to
identify these potential investments and a schedule for eliminating them from the portfolio, in
future OMB reporting.

DoD RESPONSE: DoD concurs with Recommendation 4. Current status: In future OMB
reporting, U.S. Army Corps of Engineers (USACE) will fully describe the four action plan
elements mentioned above.

RECOMMENDATION 5: To improve the U.S. Army Corps of Engineers' implementation of
PortfolioStat, we recommend that the Secretary of Defense direct the Secretary of the Army to
take the following action: Report on the agency's progress in consolidating eCPIC to a shared
service as part of the OMB integrated data collection quarterly reporting until completed.

DoD RESPONSE: DoD concurs with Recommendation 5. Current status: On 9 August 2013,
USACE obligated funds with GSA to provide eCPIC services. On 18 September 2013, GSA
completed migration of USACE's data into the eCPIC database. USACE will begin using eCPIC
for OMB reporting on or about 1 October 2013. Full implementation of eCPIC and divestiture
from the legacy Information Technology Investment Portfolio System (ITIPS) should be
complete by the end of FY14. Status updates will be provided to OMB quarterly

Appendix VIII: Comments from the Department of Energy

Department of Energy
Washington, DC 20585

September 27, 2013

Mr. David A. Powner
Director, Information Technology Management Issues
U.S. Government Accountability Office
441 G Street, NW
Washington, DC 20548

Dear Mr. Powner:

The Department of Energy's (DOE) Office of the Chief Information Officer (OCIO) appreciates the opportunity to provide comments to the General Accountability Office's (GAO) Draft Information Technology (IT) Report, *Additional OMB and Agency Actions Needed to Achieve Portfolio Savings*. We understand this audit was conducted to review the Department's implementation of PortfolioStat. DOE values the OMB PortfolioStat process and is committed to addressing the actions outlined in the report.

We offer the following Management Response to the Energy Recommendation:

Recommendation: Fully describe the following PortfolioStat action plan elements: (1) consolidate commodity IT spending under the agency CIO; and (2) establish criteria for identifying wasteful, low-value, or duplicative investments, in future OMB reporting.

The Department concurs with the recommendation.

  (1) In accordance with OMB Memorandum M-11-29, *Chief Information Officer Authorities*, dated August 8, 2011, the Department is committed to increasing the DOE CIO's oversight and authority for federal commodity IT investments through the IT Modernization Strategy issued by the Deputy Secretary in 2012. We are in the process of developing senior executive level governance boards that will make information management policy and commodity IT investment decisions for the Department. As an example, we recently established the DOE Cybersecurity Council responsible for coordinating cybersecurity planning across the Department. The Cybersecurity Council is chaired by the Deputy Secretary and involves senior level decision-makers, including the DOE CIO, from across the Department. DOE will update policy orders as necessary to implement the OMB policy and include a description in future OMB reporting.

  (2) As stated in the PortfolioStat action plan, DOE has a corporate-level budget formulation process and various IT governance boards to prioritize high-priority initiatives. The OCIO will work to establish additional value criteria to identify low-value or duplicative federal commodity IT investments. The criteria will assist the Department in identifying additional cost saving opportunities for IT that balances mission agility with efficient DOE commodity IT service delivery. The criteria will be described in future OMB reporting.

Again, thank you for the opportunity to review this report. If you have any questions related to this letter, please feel free to contact me at (202) 586-0166.

Sincerely,

Robert F. Brese

 Printed with soy ink on recycled paper

Appendix IX: Comments from the Environmental Protection Agency

UNITED STATES ENVIRONMENTAL PROTECTION AGENCY
WASHINGTON, D.C. 20460

SEP 2 5 2013

OFFICE OF
ENVIRONMENTAL INFORMATION

Mr. Paul Sabine
Assistant Director
Information Technology Management
U.S. Government Accountability Office
Washington. DC 20548

Dear Mr. Sabine:

Thank you for the opportunity to review and comment on GAO's draft report, "Information Technology – Additional OMB and Agency Actions Are Needed to Achieve Portfolio Savings." The purpose of this letter is to provide the U.S. Environmental Protection Agency's (EPA) response to your recommendations addressed to EPA.

The GAO draft report identifies a number of planned improvements outlined in OMB's PortfolioStat guidance as well as weaknesses in the implementation of PortfolioStat. The weaknesses include limitations in the CIOs' authority and in agency commodity information technology (IT) baselines, accountability for migrating selected commodity IT areas, and the information on agencies progress that OMB intends to make public. The EPA generally agrees with some of the GAO's recommendations as outlined below.

GAO Recommendation 1:
To improve the agency's implementation of PortfolioStat, we recommend that the Administrator of the Environmental Protection Agency direct the CIO to take the following three actions:
 • Develop a complete commodity IT baseline.

EPA Response:
EPA generally disagrees with this recommendation as follows:
 • A complete commodity IT baseline was provided to OMB on August 31, 2012, and uploaded to the OMB MAX site.
 • The EPA provided the following response to GAO in a letter dated April 25, 2013: "The summary commodity IT baseline information OMB provided in EPA's "PortfolioStat Agency Briefing Book" was current and complete at the time of submission in June 2012."
 • A revised commodity IT baseline and inventory was provided to the OMB MAX site on August 30, 2013, as part of the PortfolioStat submission.

GAO Recommendation 2:
 • Fully describe the following PortfolioStat action plan elements:

(1) Consolidate commodity IT spending under the agency CIO;
(2) Establish targets for commodity IT spending reductions and deadlines for meeting those targets; and
(3) Establish criteria for identifying wasteful, low-value, or duplicative investments, in future OMB reporting.

EPA Response:
(1) Consolidate commodity IT spending under the agency CIO
    EPA generally agrees with this recommendation as follows:
- The CIO is currently reviewing the commodity IT areas and is developing a plan to move existing stand alone contracts/task orders to enterprise-wide licensing agreements, where they currently do not exist today.

(2) Establish targets for commodity IT spending reductions and deadlines for meeting those targets
    EPA somewhat agrees with this recommendation as follows:
- The current review of commodity IT areas and subsequent plan development to move stand alone contracts/task order to enterprise agreements is underway. This activity will yield reductions in overall commodity IT spending through efficiencies.
- Establishing arbitrary reduction targets without corresponding requirements for IT services needed to accomplish the Agency's mission requires further evaluation.

(3) Establish criteria for identifying wasteful, low-value, or duplicative investments, in future OMB reporting
    EPA generally agrees with this recommendation as follows:
- For new investments, EPA uses its governance structure to review and assess commodity IT and other investments. The Quality and Information Council and its subcommittees, Quality Technology Subcommittee and Information Investment Subcommittee (IIS), coordinate information technology/information management and related issues.
- EPA is restructuring the IIS by broadening its scope to include greater portfolio management rigor in Capital Planning and Investment Control, Enterprise Architecture, and PortfolioStat. This will enable the IIS to establish criteria for identifying wasteful, low-value, or duplicative investments.

GAO Recommendation 3:
- Report on the agency's progress in consolidating the managed print services and strategic sourcing of end user computing to shared services as part of the OMB integrated data collection quarterly reporting until completed.

EPA Response:
The EPA generally agrees with this recommendation as follows:
- The EPA awarded a four year contract for managed print services in April of 2013. The service covers 15 geographic locations and over 1,000 printing devices. The contract vehicle has the ability to expand and service the entire EPA enterprise over time. The MPS is expected to take on additional devices and locations beginning in April 2014.

- The EPA is developing an IDIQ IT contract vehicle for purchasing and leasing of end user computing equipment. The contract is expected to be awarded by December 2013.
- The EPA is not working with other government agencies on this effort.

Thank you for the opportunity to review and comment on GAO's draft report. The EPA generally agrees with some of the GAO's recommendations and disagrees with one as outlined above. If you have any questions, please contact Fawn Freeman at 202-564-2762.

Sincerely,

Renee P. Wynn
Acting Assistant Administrator
and Acting Chief Information Officer

cc: EPA GAO Liaison Team
Fawn Freeman, Director, Mission Investment Solutions Division
Patricia Williams, OEI GAO Liaison
Anne Mangiafico, Audit Coordinator

Appendix X: Comments from the General Services Administration

The Administrator

September 25, 2013

The Honorable Gene L. Dodaro
Comptroller General of the United States
U.S. Government Accountability Office
Washington, DC 20548

Dear Mr. Dodaro:

The U.S. General Services Administration (GSA) appreciates the opportunity to review and comment on the draft report, "Information Technology: Additional OMB and Agency Actions Are Needed to Achieve Portfolio Savings (GAO-14-65)."

The U.S. Government Accountability Office (GAO) recommends that the Administrator of GSA direct their Chief Information Officer to take the following action:

- To improve the agency's implementation of PortfolioStat and report on the agency's progress in consolidating the contract writing module to a shared service as part of the OMB integrated data collection quarterly reporting until completed.

We agree with the findings and recommendations and will take action as appropriate. If you have any questions or concerns, please do not hesitate to contact me at (202) 501-0800, or Ms. Lisa Austin, Associate Administrator, Congressional and Intergovernmental Affairs, at (202) 501-0563.

Sincerely,

Dan Tangherlini
Administrator

cc: Mr. David A. Powner, Director, Information Technology Management Issues, GAO

U.S. General Services Administration
1800 F Street, NW
Washington, DC 20405
Telephone: (202) 501-0800
Fax: (202) 219-1243

Appendix XI: Comments from the Department of Homeland Security

U.S. Department of Homeland Security
Washington, DC 20528

Homeland Security

October 23, 2013

David A. Powner
Director, Information Technology Management Issues
U.S. Government Accountability Office
441 G Street, NW
Washington, DC 20548

Re: Draft Report GAO-14-65, "INFORMATION TECHNOLOGY: Additional OMB and
Agency Actions Are Needed to Achieve Portfolio Savings"

Dear Mr. Powner:

Thank you for the opportunity to review and comment on this draft report. The U.S. Department of
Homeland Security (DHS) appreciates the U.S. Government Accountability Office's (GAO's) work in
planning and conducting its review and issuing this report.

The Department is pleased to note GAO's positive acknowledgement that DHS substantially
addressed the required elements in the Fiscal Year (FY) 2012 PortfolioStat Agency Action Plan,
met required migration efforts for two commodity Information Technology (IT) areas, and
significantly contributed to PortfolioStat initiatives, cost savings, and cost avoidance.

The draft report contained one recommendation directed to DHS with which the Department
non-concurs. Specifically, GAO recommended that the Secretary of Homeland Security direct
the DHS Chief Information Officer (CIO) to:

Recommendation: Fully describe the following PortfolioStat action plan element, *consolidate
commodity IT spending under the agency CIO,* in future OMB reporting.

Response: Non-Concur. DHS has already addressed the consolidation of commodity IT
spending under the agency CIO consistent with Office of Management and Budget
Memorandum M-11-29, "Chief Information Officer Authorities," dated August 8, 2011. The
FY 2013 PortfolioStat deliverable "DHS Information Resources Management (IRM) Strategic
Plan," Appendix F, aligns the functional areas of M-11-29 to the DHS directives and delegations
that implement the functional activity. M-11-29, under Commodity IT, specifies that, "the CIO
shall pool their agency's purchasing power across their entire organization to drive down costs
and improve service for commodity IT." DHS Delegation Number 04000, "Delegation for
Information Technology," dated June 5, 2012, directs the DHS CIO to work:

- in conjunction with the DHS Chief Acquisition Officer (CAO) to eliminate duplication of commodity IT services, such as: IT infrastructure (data centers, networks, desktop computers, and mobile devices), enterprise IT systems (email, collaboration tools, identity and access management, security, and Web infrastructure), and business systems (finance, human resources, and other administrative functions).

- with the CAO and the DHS Chief Financial Officer to pool the Department's purchasing power to drive down costs and improve service for commodity IT services.

The FY 2013 DHS IRM Strategic Plan describes how the DHS CIO has been delegated the authority to oversee the Department's commodity IT spending via Delegation Number 04000, consistent with the IRM action plan element regarding consolidation of commodity IT spending under theAgency CIO. Given the foregoing explanation, we request that this recommendation be considered resolved and closed.

Again, thank you for the opportunity to review this draft report. Technical comments were provided under separate cover. Please feel free to contact me if you have any questions. We look forward to working with you in the future.

Sincerely,

Jim H. Crumpacker
Director
Departmental GAO-OIG Liaison Office

2

Appendix XII: Comments from the Department of Housing and Urban Development

U.S. DEPARTMENT OF HOUSING AND URBAN DEVELOPMENT
WASHINGTON, DC 20410-3000

CHIEF INFORMATION OFFICER

SEP 2 3 2013

Mr. David A. Powner
Director
Information Technology Management Issues
U.S. Government Accountability Office
441 G Street NW
Washington, DC 20548

Dear Mr. Powner:

Thank you for the opportunity to comment on the Government Accountability Office (GAO) draft report entitled, *Information Technology: Additional OMB and Agency Actions Are Needed to Achieve Portfolio Savings* (GAO-14-65).

The Department of Housing and Urban Development (HUD) reviewed the draft report and concurs with the recommendations for Executive Action. More definitive information with timelines will be provided once the final report has been issued. HUD remains committed to meeting the Office of Management and Budget's (OMB) PortfolioStat requirements.

If you have any questions or require additional information, please contact Joyce M. Little, Chief, Audit Compliance Branch, at (Joyce.M.Little@hud.gov), or 202-402-7404.

Sincerely,

Barbara A. Elliott
Acting Chief Information Officer

Appendix XIII: Comments from the National Aeronautics and Space Administration

National Aeronautics and Space Administration

Headquarters
Washington, DC 20546-0001

SEP 2 7 2013

Reply to Attn of: Office of the Chief Information Officer

Mr. David Powner
Director
Information Technology Management Issues
United States Government Accountability Office
Washington, DC 20548

Dear Mr. Powner:

The National Aeronautics and Space Administration (NASA) appreciates the opportunity to review and comment on the Government Accountability Office (GAO) draft report entitled, "Information Technology: Additional OMB and Agency Actions Are Needed to Achieve Portfolio Savings" (GAO-14-65).

In the draft report, GAO addresses three recommendations to the NASA Administrator. To improve implementation of the Office of Management and Budget (OMB) PortfolioStat process, GAO recommends that the NASA Administrator direct the Chief Information Officer to take the following actions:

Recommendation 1: Reflect 100 percent of information technology investments in the agency's enterprise architecture.

Management's Response: Concur. NASA continues to work toward reflecting all information technology investments in the enterprise architecture. NASA's policy regarding Enterprise Architecture (NPR 2830.1) is currently under review and describes the architecture process that will help enable this. NASA is also working with the OMB Federal Enterprise Architect to meet the requirements as defined in the Common Approach to Federal Enterprise Architecture. High-level target architectures have been established for all of the core IT infrastructure domains (applications, compute, communications, end user, information, and security) and roadmaps have been developed for each that identify the transition investments over the next five years. NASA IT research and development investments have been described in the Strategic Space Technology Investment Plan and its associated IT Roadmaps. An Enterprise Architecture Board charter is being developed that will enable the board to oversee architecture activities and ensure IT investments are included in the architecture. An architecture checklist has been developed and is being utilized to help ensure investments align with the architecture. NASA is also in the early stages of

2

implementing an enterprise architecture tool that will allow for more easy inclusion of investments into the architecture.

NASA will work toward reaching a goal of 50 percent by the end of FY14 and 90 percent by the end of FY15. NASA will strive to reflect 100 percent of information technology investments in the Agency's enterprise architecture; however, due to NASA's evolving mission and the reality of Government budget cuts, the 100 percent target may not be attainable.

Recommendation 2: Fully describe the following PortfolioStat action plan elements: (1) consolidate commodity IT spending under the agency CIO; (2) target duplicative systems or contracts that support common business functions for consolidation; (3) establish criteria for identifying wasteful, low-value, or duplicative investments; and (4) establish a process to identify these potential investments and a schedule for eliminating them from the portfolio, in future OMB reporting.

Management's Response: Concur. Over the past ten years, NASA has consolidated a significant number of commodity IT systems and services and implemented an Agency shared-services center. As an example, the Integrated Enterprise Management Program (2000-2009) consolidated multiple Center Financial Systems into one Agency-wide Core Financial Management system and Business Warehouse; consolidated Real Property and Personal Property Management systems and integrated those with the financial system; and consolidated and integrated Human Capital Systems, integrated data into Business Warehouse, and interfaced all HR systems. Further, NASA launched the NASA Shared Services Center (NSSC) in 2006 to perform Agency-wide transactional and administrative activities that had been performed at each of our Centers.

Additionally, NASA's Information Technology Infrastructure Integration Program (I3P) (2007-Present) has consolidated NASA's e-mail services into one system; consolidated IT security incident management and response across the Agency through the Security Operations Center (SOC); centralized IT contract and financial management to support the Agency I3P program; established an Agency Enterprise Service Desk as an integrated multi-tier support for service delivery; and launched Agency Consolidated End-User Services (ACES) to maintain the end-user IT inventory. In 2012, the Agency established the NASA Integrated Communications Services (NICS) to consolidate and operate NASA's communication capabilities and to provide enterprise management services. In 2013, an Agency Web Services contract (WESTPrime) was awarded to utilize cloud services and to use open source software for Web site and Web application development and hosting, which will result in additional savings.

NASA will continue to fully describe these and future efforts in the OMB quarterly PortfolioStat reporting updates. NASA estimates completion by June 2014.

3

Recommendation 3: Report on the agency's progress in consolidating the NASA Integrated Communications Services Consolidated Configuration Management System (NC2MS) to an enterprise-wide shared service as part of the OMB integrated data collection quarterly reporting until completed.

Management's Response: Concur. NASA will report on the progress of NC2MS until the project is complete, which is currently scheduled for February 2014.

Thank you for the opportunity to comment on this draft report. If you have any questions or require additional information, please contact Valarie Burks at (202) 358-3716.

Larry Sweet
Chief Information Officer

NATIONAL
ARCHIVES

ARCHIVIST of the
UNITED STATES

DAVID S. FERRIERO
t: 202.357.5900
f: 202.357.5901
david.ferriero@nara.gov

24 September 2013

David A. Powner
Director, Information Technology Management Issues
United States Government Accountability Office
44 G Street, NW
Washington, DC 20548

Dear Mr. Powner:

Thank you for the opportunity to review and comment on the Government Accountability Office's (GAO's) draft report 14-65 titled "*Additional OMB and Agency Actions Are Needed to Achieve Portfolio Savings*." We concur with your recommendation to fully describe several elements in our PortfolioStat action plan, including:

1. Consolidate commodity IT spending under the Agency CIO;
2. Target duplicative systems or contracts that support common business functions for consolidation;
3. Establish criteria for identifying wasteful, low-value, or duplicative investments; and
4. Establish a process to identify these potential investments and a schedule for eliminating them from the portfolio.

We will include new or updated descriptions in future OMB reporting.

If you have any questions regarding this memo, please contact Carla Riner, Deputy Chief Operating Officer, at 301-837-0643 or via email at carla.riner@nara.gov.

Sincerely,

DAVID S. FERRIERO
Archivist of the United States

Via email to: David A. Powner, pownerD@gao.gov
                Sabine R. Paul, PaulS@gao.gov

NATIONAL ARCHIVES and
RECORDS ADMINISTRATION

700 PENNSYLVANIA AVENUE NW
WASHINGTON, DC 20408-0001
www.archives.gov

Appendix XV: Comments from the National Science Foundation

NATIONAL SCIENCE FOUNDATION
4201 WILSON BOULEVARD
ARLINGTON, VIRGINIA 22230

Mr. David A. Powner
Director, Information Technology Management Issues
U.S. Government Accountability Office
441 G Street, NW
Washington, DC 20548

Dear Mr. Powner:

Thank you for the opportunity to provide comments on the draft GAO Report "Information Technology: Additional OMB and Agency Actions are Needed to Achieve Portfolio Savings" (GAO-14-65).

NSF is pleased to note GAO's generally positive assessment of our compliance with PortfolioStat initiative requirements. NSF generally agrees with GAO's characterization of our status, but wishes to provide clarification about the completeness of our PortfolioStat Action Plan.

GAO's assessment indicates that two elements of NSF's PortfolioStat Action Plan were not complete. NSF believes that the agency PortfolioStat Action Plan does include information about the agency's activities to consolidate commodity IT spending under the agency CIO and to establish criteria for identifying duplicative, low-value, and wasteful investments. Per GAO's recommendation, NSF will plan to update the PortfolioStat Action Plan as appropriate to more fully describe these activities.

NSF also wishes to provide an update on the status of agency commodity IT migration efforts. NSF's cloud email migration efforts were successfully completed in August 2013. We reported the status of our efforts to OMB during our PortfolioStat session, and will follow up with formal reporting as required.

Again, thank you for the opportunity to review and comment on this draft report. We look forward to working with you on future NSF engagements. If you have any questions or concerns, please feel free to contact me at (703) 292-8100.

Sincerely,

for Amy Northcutt
Chief Information Officer

Appendix XVI: Comments from the Office of Personnel Management

UNITED STATES OFFICE OF PERSONNEL MANAGEMENT
Washington, DC 20415

Chief Information
Officer

September 23, 2013

David A. Powner
Director, Information Technology Management Issues
Government Accountability Office
441 G. ST. N.W.
Washington, D.C. 20548

OPM Responses to Draft GAO Report GAO-14-65

The U.S. Office of Personnel Management (OPM) has reviewed the Government Accountability Office (GAO) draft audit report (GAO-14-65) on OPM's PortfolioStat program and is in concurrence with the findings and recommendations identified in the report. We recognize that the OPM programs can benefit from an external evaluation and we appreciate the GAO input as we continue to work to enhance the OPM PortfolioStat Review process. Specific responses to your recommendations are provided below.

GAO Recommendation 1: Develop a complete commodity IT baseline.

OPM Management Response: OPM concurs with the GAO recommendation. OPM had submitted a commodity IT baseline to OMB on June 15, 2012. OPM will continue to update the commodity IT baseline as required by OMB on November 30, 2013.

GAO Recommendation 2: Fully describe the following PortfolioStat action plan elements: *(1) move at least two commodity IT areas to shared services; and (2) target duplicative systems or contracts that support common business functions for consolidation,* in future OMB reporting.

OPM Management Response: OPM concurs with the GAO recommendation. OPM has completed several elements and continued to plan for the two shared services initiatives and will provide the planning elements as required by the next OMB reporting due November 30, 2013. OPM is still performing on-going review of contracting processes to address the consolidation issues and will provide information to the next OMB reporting due November 30, 2013.

GAO Recommendation 3: Report on the agency's progress in consolidating the help desk consolidation and IT asset inventory to shared services as part of the OMB integrated data collection quarterly reporting until completed.

OPM Management Response: OPM concurs with the GAO recommendation. OPM has completed some of the help desk consolidation elements and continues to make strides in future planning for additional consolidation. OPM is continuing to develop the IT asset inventory shared service offering. OPM will provide progress information to the OMB integrated data collection quarterly reporting as required by OMB on November 30, 2013.

Please contact Ms. Janet Barnes at (202) 606-3207 should your office require additional information.

Sincerely,

Charles Simpson
Acting, Chief Information Officer
U.S. Office of Personnel Management

Appendix XVII: Comments from the Social Security Administration

SOCIAL SECURITY
Office of the Commissioner

September 27, 2013

Mr. David A. Powner
Director, Information Technology Management Issues
United States Government Accountability Office
441 G. Street, NW
Washington, DC 20548

Dear Mr. Powner,

Thank you for the opportunity to review the draft report, "INFORMATION TECHNOLOGY:
Additional OMB and Agency Actions Are Needed to Achieve Portfolio Savings." Our response
to the audit report contents, findings, and recommendations are enclosed.

If you have any questions, please contact me at (410) 965-0520. Your staff may contact
Gary S. Hatcher, Senior Advisor for Records Management and Audit Liaison Staff, at
(410) 965-0680.

Sincerely,

Katherine Thornton
Deputy Chief of Staff

Enclosure

SOCIAL SECURITY ADMINISTRATION   BALTIMORE, MD 21235-0001

**COMMENTS ON THE GOVERNMENT ACCOUNTABILITY OFFICE DRAFT
REPORT, "INFORMATION TECHNOLOGY: ADDITIONAL OMB AND AGENCY
ACTIONS ARE NEEDED TO ACHIEVE PORTFOLIO SAVINGS" (GAO-14-65)**

We remain committed to PortfolioStat objectives and our complete and accurate compliance of
its requirements.

Recommendation 1

Develop a complete commodity Information Technology (IT) baseline.

Response

We do not agree with your assessment that our commodity Information Technology (IT) baseline
is incomplete. We believe our commodity IT baseline data are complete and accurate. While we
did not develop a separate formal process to comply with PortfolioStat, our baseline is a subset,
derived through a data call, of our rigorous IT Capital Planning and Investment Control Process.
We will formally document our process to demonstrate the completeness of our commodity IT
baseline in our next annual cycle.

Recommendation 2

Report on the progress in consolidating the geospatial architecture to a shared service as part of
the Office of Management and Budget (OMB) integrated data collection quarterly reporting until
completed.

Response

We agree. The two shared service efforts we designated for migration were Enterprise Social
Media and Geospatial Architecture (GA). We completed migration of our Enterprise Social
Media in December 2012. For GA, we established a work group to develop a business case and
Geospatial Information Systems architecture, which will provide a common resource for
business users across the agency. We have completed a number of activities that included
implementing a primary infrastructure, implemented our shared-license strategy, conducted a
supporting connection to test and develop a best practice environment. However, deployment
and testing with existing product users has extended the timeline for full consolidation and
access capability for all users. We expect to fully implement GA by September 2014.

Appendix XIII: Comments from the Department of State

United States Department of State
Comptroller
P.O. Box 150008
Charleston, SC 29415-5008

SEP 25 2013

Dr. Loren Yager
Managing Director
International Affairs and Trade
Government Accountability Office
441 G Street, N.W.
Washington, D.C. 20548-0001

Dear Dr. Yager:

We appreciate the opportunity to review your draft report, "INFORMATION TECHNOLOGY: Additional OMB and Agency Actions Are Needed to Achieve Portfolio Savings" GAO Job Code 311280.

The enclosed Department of State comments are provided for incorporation with this letter as an appendix to the final report.

If you have any questions concerning this response, please contact Colleen Hinton, IT Manager, Bureau of Information Resource Management at (202) 634-0320.

Sincerely,

James L. Millette

cc: GAO – David A. Powner
IRM – Steven Taylor
State/OIG – Norman Brown

Department of State Comments on GAO Draft Report

INFORMATION TECHNOLOGY: Additional OMB and Agency Actions are
Needed to Achieve Portfolio Savings
(GAO-14-65, GAO Code 311280)

Thank you for the opportunity to comment on your draft report entitled
*"Information Technology: Additional OMB and Agency Actions are Needed to
Achieve Portfolio Savings."*

The GAO draft report provides three recommendations for the Department
to improve the implementation of FY12 PortfolioStat actions: (1) ensure that
100% of IT investments are in the Department's enterprise architecture, (2) update
FY12 PortfolioStat Action Plan to address additional elements, and (3) report to on
the progress of consolidating the Foreign Affairs Network and content publishing
and delivery services as part of OMB's Integrated Data Collection process. In
response to GAO's recommendations, the Department concurs with comment;
State would like to clarify to GAO that State has already consolidated more than
two commodity IT areas per OMB Memorandum M-11-29. Specifically, State has
consolidated systems for email, human resources, financial management,
procurements, and a consolidated IT refresh supply chain service. The Department
will develop responses to each of the GAO recommendations once the report is
published, as well as update OMB's Integrated Data Collection to reflect the status
of current activities.

Appendix XIX: Comments from the Department of Veterans Affairs

DEPARTMENT OF VETERANS AFFAIRS
Washington DC 20420

September 30, 2013

Ms. Debra A. Draper
Director, Health Care
U.S. Government Accountability Office
441 G Street, NW
Washington, DC 20548

Dear Ms. Draper:

The Department of Veterans Affairs (VA) has reviewed the Government Accountability Office's (GAO) draft report, *"INFORMATION TECHNOLOGY: Additional OMB and Agency Actions Are Needed to Achieve Portfolio Savings"* (GAO-14-65). VA generally agrees with GAO's conclusions and concurs with GAO's four recommendations to the Department.

VA is taking concrete steps to manage its investment portfolio more effectively. The actions identified in the attachment will help to foster an environment at VA in which investments can be evaluated with respect to their value to the enterprise as a whole. VA will be able to assess the extent to which a planned investment either improves mission execution or decreases the cost of operations. Further, the Department will be able to measure how well the investment meets goals throughout its lifecycle.

The enclosure specifically addresses GAO's four recommendations and provides an action plan for each. VA appreciates the opportunity to comment on your draft report.

Sincerely,

Jose D. Riojas
Chief of Staff

Enclosure

Enclosure

Department of Veterans Affairs (VA) Response to
Government Accountability Office (GAO) Draft Report
*"INFORMATION TECHNOLOGY: Additional OMB and Agency Actions Are
Needed to Achieve Portfolio Savings"*
(GAO-14-65)

**To improve the department's implementation of PortfolioStat, we recommend that
the Secretary of Veterans Affairs direct the CIO to take the following four actions:**

**GAO Recommendation 1: Fully describe the following PortfolioStat action plan
element, *target duplicative systems or contracts that support common business
functions for consolidation,* in future OMB reporting.**

VA Comment: Concur. VA's Office of Information Technology's (OIT) Ruthless
Reduction Task Force (RRTF) continues to identify opportunities for cost avoidance in
VA information technology (IT) expenses by a variety of means including
decommissioning redundant systems; consolidation of development and test
environments; implementation of service-oriented architecture, data center
consolidation, and cloud computing; and efficiencies in obtaining security approvals for
would-be employees and contractors.

The OIT Vendor Management Office (VMO) was established to provide a strategy map
for vendor management and to clearly define roles and responsibilities across the
enterprise. OIT VMO works hand-in-hand with the RRTF to identify duplicative efforts,
cost avoidances with enterprise license agreements negotiations, and strategic sourcing
opportunities.

Future Office of Management and Budget (OMB) Reporting

RRTF recommendations will include activities that contribute to the improvement of
information resource management in a sustainable manner and will be used to inform
the annual budget formulation process. Task force findings will also be reported in
accordance with the OMB Fiscal Year (FY) 2013 PortfolioStat Guidance (M-13-09) on a
quarterly basis.

- VA is actively introducing Enterprise Architecture (EA) products and information into
  the strategic planning, programming, and budgeting processes within the
  Department as a whole and OIT in particular. VA recognizes EA's role in providing
  critical information to support business process integration, enterprise data
  governance, application architecture and service-oriented architecture, and
  infrastructure rationalization.
- In particular, VA is identifying opportunities to leverage EA information in our RRTF
  work and use RRTF findings in our systems engineering and investment decision
  processes.
- VA is strengthening the Department-wide IT Planning, Programming, Budgeting and
  Execution (PPBE) process which will emphasize the use of alternative analyses and
  business cases as input to key decision processes. VA's IT investment

1

Enclosure

Department of Veterans Affairs (VA) Response to
Government Accountability Office (GAO) Draft Report
*"Information Technology: Additional OMB and Agency Actions Are Needed to
Achieve Portfolio Savings"*
(GAO-14-65)

management processes will align to the Department's ensuring that consistent
criteria and analyses are used at all phases of portfolio management.

**GAO Recommendation 2: Report on the department's progress in consolidating
the dedicated fax servers to a shared service as part of the OMB integrated data
collection quarterly reporting until completed.**

VA Comment: Concur. VA did not accomplish Elimination of Analog Fax Lines in
2012. VA is not stepping back from the cost containment idea, but it did not accomplish
the effort as planned. When the cost basis was revised by leadership, it was
determined that the return on investment (ROI) fell to 0.55 from 7.59. As a result, this
effort is being reprioritized and considered for the FY 2014 budget cycle.

**GAO Recommendation 3: As the department matures its enterprise architecture
process, make use of it as well as the valuation model to identify whether there
are additional opportunities to reduce duplicative, low-value, or wasteful
investments.**

VA Comment: Concur. VA is working both to mature its EA and to incorporate
architectural products and information into key enterprise processes. In 2012, the
OneVA EA program developed Compliance Criteria as a guide to the Department's
Enterprise Technical Architecture; these criteria are used in assessments of projects in
connection with VA's Project Management Accountability System, the governance
structure for IT systems development.

Members of the OneVA EA team have been collaborating with the Director of Strategy
for VA and are now working with the office in charge of implementing the Department's
updated PPBE processes. The goal of the EA team is to develop architecture products
that will inform these processes and promote the use of architecture as the decision
support tool to inform key Department decisions, particularly in the planning and
programming cycles. These products (views, reports, and analyses) will be used to
identify redundancies as well as gaps in planned investments.

The EA team will be working with members of the RRTF to familiarize them with EA and
to make EA products available to the RRTF in their analyses. Members of the team are
also beginning work with the Information Technology Resources Management
organization within OIT in order to 1) develop the appropriate alignment of budget
information to architectural information and 2) use EA (gap and redundancy analyses,
etc.) to inform the IT budget formulation process. The results of these collaboration
efforts will be incorporated into scheduled EA releases.

2

Enclosure

Department of Veterans Affairs (VA) Response to
Government Accountability Office (GAO) Draft Report
*"Information Technology: Additional OMB and Agency Actions Are Needed to
Achieve Portfolio Savings"*
(GAO-14-65)

By the end of FY 2014, VA expects to have EA products in use and informing PPBE
decision processes and IT budget formulation processes.

**GAO Recommendation 4: Develop detailed support for the estimated savings for
fiscal years 2013 to 2015 for the Server Virtualization, Eliminate Dedicated Fax
Servers Consolidation, Renegotiate Microsoft Enterprise License Agreement, one
CPU policy initiatives.**

<u>VA Comment:</u> Concur. Information on specific topics is provided below:

Virtualize Server: Initiative was initially scoped to achieve 50 percent server
virtualization with anticipated cost avoidance (reported in the PortfolioStat as of
August 2013) as follows: in FY 2012 – $84,000; FY 2013 – $424,000; FY 2014 –
$754,000; and FY 2015 – $1.1 million. During FY 2013, the initiative was re-scoped to
increase server virtualization to 75 percent. The RRTF team is reworking the ROI and
cost avoidance estimates based on the revised scope.

Eliminate Dedicated Fax Servers: See current status detailed under recommendation 2.

Renegotiate Microsoft Enterprise License Agreement: Total savings over the 5-year
period renegotiated is $161 million. Annual costs were reduced from $114 million per
year to $77 million, with a one-time "true up" fee of $24 million paid in FY 2012.

One CPU Policy: To maximize resource utilization, reduce VA's IT footprint, and
eliminate inefficiencies (and potential conflicts and confusion) due to competing policies,
the One CPU Policy and Field Mobile Workers and Tele-work Support Agreement
(i.e., End User Device Support Agreement) were integrated into one strategy. The End
User Device Support Agreement will provide guidance and expectations to standardize
IT devices across the enterprise while promoting telework. A new economic justification
is being developed based on the consolidated strategy and will be provided when
completed.

3

Appendix XX: GAO Contact and Staff Acknowledgments

GAO Contact	David A. Powner, (202) 512-9286 or pownerd@gao.gov
Staff Acknowledgments	In addition to the contact named above, individuals making contributions to this report included Sabine Paul (Assistant Director), Valerie Hopkins, Lee McCracken, Tomas Ramirez, and Bradley Roach.